W9-BNL-060

# POISONED NATION

ALSO BY LORETTA SCHWARTZ-NOBEL

Nonfiction

*Starving in the Shadow of Plenty*
*Engaged to Murder*
*A Mother's Story*
*The Baby Swap Conspiracy*
*Forsaking All Others*
*Growing Up Empty*

Fiction

*The Journey*

# POISONED NATION

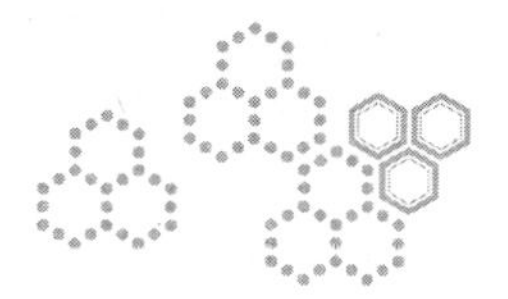

## Pollution, Greed, and the Rise of Deadly Epidemics



LORETTA SCHWARTZ-NOBEL

ST. MARTIN'S PRESS NEW YORK

 For information, address St. Martin's Press, 175 Fifth Avenue, New York, N.Y. 10010.

www.stmartins.com

Design by Susan Walsh

Library of Congress Cataloging-in-Publication Data

Schwartz-Nobel, Loretta.
Poisoned nation: pollution, greed, and the rise of deadly epidemics / Loretta Schwartz-Nobel.—1st ed.
p. cm.
ISBN-13: 978-0-312-32797-2
ISBN-10: 0-312-32797-8
1. Environmental toxicology—United States. 2. Pollution—Environmental aspects—United States. 3. Epidemics—United States. 4. Corporations—Corrupt practices—United States.

RA1226.S39 2007
615.9'02—dc22 2007015356

First Edition: August 2007

10 9 8 7 6 5 4 3 2 1

*For my sister Helen Driscoll,*
*who died on July 28, 2006,*
*of a rare environmentally linked adrenal cancer,*
*just three weeks after diagnosis*

# CONTENTS

# ACKNOWLEDGMENTS

I AM A JOURNALIST WHO HAS RELUCTANTLY TAKEN ON A VERY large task. I could never have done it alone. Although the responsibility for the text is solely mine, the intellect and scholarship of countless environmentalists, researchers, activists, authors, scientists, journalists, physicians, politicians, and religious leaders inform these chapters. Investigative reporters are a little like beachcombers. We walk life's shores endlessly searching beneath the sand and waves. We cull and dig and collect, then polish and string together the fragments and pearls of those who have come before us. We interpret, synthesize, and try to create from all those fragments a larger image, a bigger picture, a new vision.

Among the environmentalists and journalists to whom I am most indebted are Rachel Carson, Robert Kennedy, Jr., Renee Sharp, Bill Walker, Nancy Evans, Bill Moyers, Glenn Scherer, Marla Cone, Sarah Ruby, Sandra Steingraber, Ross Gelbspan, Jeffrey St. Clair, Renee Downing, Peter Waldman, Erica Werner, Michael Bender, Colleen Diskin, Betty Brink, Lindy Washburn, Alex Nessbaum, Bob Feldman, Miguel Bustillo, Peter Eisler, Kurt Gottfried, Deborah Frisch, and Al Gore, whose *Inconvenient Truth* helped me to make the connections between human illness, environmental contamination, and global warming.

Among the physicians and scientists whose research guided me are Dr. Samuel Epstein, Dr. Eric Dewailly, Dr. Russell L. Blaylock, Dr. John Grofman, Dr. Amy Holmes, Dr. Tim O'Shea, and Dr. Janette D. Sherman.

Among the politicians who have worked for change are Senators Dianne Feinstein, Hillary Rodham Clinton, Ted Kennedy, Harry Reid, and Jim Jeffords.

Among the great religious leaders and groups who have voiced their profound concern are Pope John Paul II, Ecumenical Patriarch Bartholomew, the National Religious Partnership for the Environment, the Interfaith Council of Churches, Rabbi Saul Berman, Mark Jacobs, Paul Gorman, Dr. Robert Edgar, John Carr, Dr. John Ruskay, Rev. Dr. Joan Brown Campbell, Dr. Ronald Sider, and Fr. Chris Bender.

Among the activists who have inspired me are Barbara Brenner, Barbara Loe Fisher, Andrea Ravinett Martin, and Larry Ladd.

Among the victims of chemical contamination and their families, whose stories have moved me to search further, are Neil and Karen Blair, Ellen Harris, Jerry Ensminger and his daughter Janey, Belinda Reanda, Virginia Morales, her daughter, Desiree Hoguin, and grandson Aaron, Kathleen and Kristen Froess, Anastacia and Matt Warnecke, Adam Jernee, and Greg Voetsch and his entire family.

I also want to thank my former husband, Joel Nobel, and my good friends Mark Griffis, David Robertson, Gay Heib, Judy Talbert, Joe and Bonnie Ryan, Maryellen Wheeler, Bo Dean, Cyndi and Ron McNeil, Pam and Jack Robillard, Katy and John Bridge, Barbara Brigman, Carol Saline, Yoko Johanning, Jeffry Burr, Neil Blair, Fr. Alexis Regis, and Rosemary Gross for their untiring support.

This book could never have been written without the faith of my editors, Diane Reverand and Phil Revzin, and production editor, Julie Gutin; the tough professional guidance of my agent, Ellen

Levine; the assistance of my late sister Helen Driscoll, who frequently provided me with invaluable material and ideas; the skillful editorial and literary suggestions of my daughter Ruth, who helped with an earlier draft of the manuscript; the enthusiasm and support of my other children, Rebecca and Adam; the inspiration of my new grandson, Jonathan; the treasured memory of my mother and father, Fay and Abraham Rosenberg; the word-processing skill, editorial assistance, love, and endless patience of my partner, Bill Burr; and, finally, a guiding spiritual force and presence far larger than anything that I can comprehend.

# PREFACE

## Stolen Lives:
## A Personal Awakening

> Children of God—look at My works! See how beautiful they are . . . do not spoil and destroy My world, for if you do, there will be no one else to repair it.
>
> —MIDRASH ECCLESIASTES RABBAH 7:13

WHEN MY EDITOR DIANE REVERAND CALLED AND SAID SHE WAS concerned about the epidemic surges in autism, birth defects, breast cancer, asthma, and clusters of other illnesses in cities and towns around the United States, it began a chain of research that opened my eyes to a frightening world of hidden contamination around the globe that threatens all of us.

It was the 9/11 era, and I was still more concerned with terrorists and anthrax than with the deadly contamination that America and the rest of the world were bringing upon themselves. I still felt perfectly safe when I breathed the air around me or drank water from the tap. When I drank milk or ate salad, fruit, vegetables, chicken, or fish, I thought more about avoiding carbohydrates and calories than about avoiding chlorinated hydrocarbons, perchlorate, arsenic, mercury, and thousands of other untested and often toxic chemicals. I still sprayed household pests with insecticides, rubbed mosquito repellent on my skin and my children's skin, and used cosmetics without concern. I still cleaned my floors and sprayed my shower curtains with whatever chemical cleansers happened to have accumulated in the cabinet under my kitchen sink.

I had recently moved from Philadelphia, Pennsylvania, to a relatively remote barrier island on the coast of North Carolina. Even though it was only about ten miles from the Camp Lejeune military base and forty-five miles from the small, bustling city of Wilmington, I felt safe from the man-made contaminants and pollution of the big cities.

A house built right on the dunes at the edge of the ocean was the dream of my lifetime and my mother's before me. After her death, I believed it was then or never, so I searched the entire East Coast for something remote enough to be affordable. Then, when I found it, I looked up at the stars and clouds and sensed that my mother knew we were finally there. I thought it was the purest and most primitive place I had ever lived. Countless pelicans and seagulls flew overhead. Giant sea turtles nested just two hundred feet away from my door, and sand crabs scurried into holes as I took my evening walks. Sometimes, when the moon was full and the tide was high and the ocean came right to the foot of my deck, a fear of the encroaching sea or the threat of hurricanes engulfed me, but a fear of environmental contamination never did. Ironically, just as my life overflowed with the joy of my isolated fulfillment, the contamination that threatens all of us was moving ever closer.

Day after day, as I sat at my desk writing and looking out at the changing tides and the schools of dolphins, I noticed the warships that dotted the distant horizon, but I knew little about the sailors and Marines from nearby Camp Lejeune who manned those ships. I had not yet heard that thousands of military men, women, and children had been sick and dying there for years while continuing to drink water that was contaminated. Not until months later did I learn that "the number of people who drank PCE- and TCE-tainted water, bathed in it, and had water fights with it is staggering. The Marine Corps estimates that 50,000 Marines and their families lived in the base housing that had been fed by the wells before they were closed. . . . Victim advocacy groups place the figure even higher at 200,000, which would make Camp Lejeune one of the largest contaminated-water cases in U.S. history. Already more than 270 tort claims have been filed with the Navy's judge advocate general's office by former residents."[1]

Nor did I know that my own drinking water was soon to be merged by public agreement with the water from Camp Lejeune. I was joyfully ignorant, captivated by my new sense of reverence for the natural order of all things, for the cycles of life that had evolved over millions and millions of years, and for the amazing synergism and balance that clearly existed between living things and their surroundings.

It was not until several months after the call from my editor suggesting that I write a book about illness and environmental contamination that I discovered the leukemia, birth defects, and devastating cluster illnesses of my neighbors at Camp Lejeune. After that, the book evolved and seemed to take on a life of its own. I soon learned that a chemical called perchlorate, used to fuel Cold War–era missiles and the rockets that put men on the moon, now polluted much of the Colorado River, which was the water source for more than 30 million people across the Southwest. Then I found out that perchlorate had also invaded the water and food supply of millions of other people in cities and towns scattered throughout the nation. This prompted me to begin a more comprehensive investigation of contaminated food, air, and everyday products, and of the dangerous bonds that often exist between science, industry, and the government. The horrifying truths about the autism cover-up, the cancer industry, and the new strategies for deception that have led to a global lung cancer pandemic soon followed. There were days when the research left me numb with despair, days when anger and outrage spurred me forward, and days when only my faith in something larger and wiser than ourselves kept me going. The despair and anger I experienced was often shared by the environmental groups and women's advocates that I came to know, but it was always matched by their restraint and their determination to be wiser and more prudent than the people whose policies they fought against. They never engaged in acts of retribution. They never followed the violent policies of a small handful of environmental extremists who had given the environmental movement a bad name.

As the book evolved, I found myself inspired by the people I interviewed. My spiritual faith grew. Ultimately, so did my optimism. I slowly became convinced that religious groups working together had the power, the obligation, and the opportunity to bring about change, to care for creation, and to save the earth for future generations.

Of course, I had known that human beings have acquired the ability to change the natural balance of things for better or worse in the last hundred years. I knew we could alter basic chemistries in our laboratories, factories, power plants, and military bases in ways that even the most advanced of our scientists could only partly understand or predict. For the most part, I had accepted the slogan "Better living through chemistry" without really questioning it. I had heard that many of our synthetic creations have no counterparts in nature, but I didn't fully understand that a chain of deadly contamination had already been distributed throughout the world simply for profit, or that the contamination occurred virtually everywhere—in the air we breathe, the water we drink, the food we eat, the vaccines our children are given, the makeup we wear, and the other products we use every day. I did not realize the extent to which dangerous poisons are also in the streams flowing deep beneath the earth, in the soil used by our farmers for planting crops, and in the water used to irrigate them. Nor did I understand that toxic chemicals are in the bodies of the fish, chickens, cattle, and pigs and the fruits, vegetables, and rice we eat as well as in the bodies of other domestic and wild animals. I did not realize that hundreds of untested chemicals have already infiltrated each and every one of our own bodies, the bodies of our growing children, our newborn infants, and even our unborns.

I didn't know these things, at least in part because many politicians, chemical companies, and pharmaceutical companies didn't want me or other mothers like me to know that, completely against our will, our maternal instinct had been violated. We have become a species that violates life's most basic principles, a species that poisons our own infants while they are still in our wombs through the permeable barrier of our placenta and then continues the process by nursing them with our contaminated breast milk. That which sustains life and that which destroys life have been steadily merging for many years.

I still believed that the Environmental Protection Agency, the Food and Drug Administration, the Centers for Disease Control and Prevention, the American Pediatric Association, the American Cancer Society, the National Cancer Institute, and the American Academy of Sciences were pillars of authority and could offer the final, truthful word in each of their respective fields. I did not yet know about the intense pressure to alter or suppress information to which these groups and others are sometimes subjected, or about the partnerships and sustaining bonds that too often link some of our most powerful politicians, scientific societies, drug companies, and other profit-making corporations.

I had just finished writing a book about hunger and poverty in America, and I was still more concerned with children who didn't have enough to eat than with children who ate food, drank water, and breathed air that would damage all and slowly kill many of them. I did not yet know that 90 percent of childhood leukemias and as many as 70 percent of breast cancers are now believed to be caused by man-made environmental contaminants.

In the months that followed that path-changing call from my

editor, my own life was to undergo a radical transformation and be darkened by the staggering new reality. I was finally to understand that what separates man-made poisons from those derived from nature is the conscious, often arrogant and careless manipulation of chemicals by scientists who frequently do not know what the final outcome will be. Whether their motive is corporate profit, personal greed, or political power, the outcome is often uncontrollable and deadly.

When combinations of these chemicals enter our bodies and settle in our blood, our brains, our flesh, and the deepest marrow of our bones, they can prevent the normal functioning of our organs and initiate the slow but irreversible changes that ultimately lead to cluster illnesses, birth defects, autism, malignancies, and other illnesses.

I didn't yet realize that a day-by-day buildup of lethal chemicals is taking place in all of our bodies, all of the time, or that, as Rachel Carson pointed out in her book *Silent Spring,* our constant exposure to multiple poisons, however slight and incremental, could trigger profound neurological changes and result in slow, cumulative destruction. I had no idea that we are eating, drinking, breathing, and using seemingly harmless materials every single day that, taken together, could turn out to be fatal.

Nor did I understand that global warming follows a parallel path, that the planet itself is endangered by the same sources of abuse. I did not yet know that man's reckless misuse of his technical gifts simultaneously threatens the earth and its children.

In both cases, what even the best-intentioned experts tell us about "safe" levels of deadly chemicals often turns out to be wrong, because they, too, are learning more as time goes on. Scientists are continually forced to revise their estimates of safety as new information becomes available.

Most of the estimated 85,000 synthetic chemicals found in the United States and other parts of the world today have never been tested for their destructive effects. They have simply been sent into our world and have become an unavoidable part of our everyday lives. Of those that have been studied, some are now absolutely known to cause global warming or disease. Among the known carcinogens, for example, there are some—polychlorinated biphenals (PCBs) and DDT—that were banned decades ago but will remain in the environment for many years to come. There are others known to be toxic, like perchlorate, mercury, and arsenic, that continue to be used and argued over while we and our children regularly consume or breathe them.

Whether it is contamination from air, water, food, or the everyday things we buy and use, the patterns are the same. We are surrounded by life-threatening contaminants and carcinogens that we cannot avoid or control. Most of us have relatively little knowledge of the threat and, of course, we had nothing to do with making the decisions that have set these chemical events in motion. We were not protected by those in authority, and we were not consulted. In fact, we were often deliberately misinformed. What science conceives, industry makes possible. Government and big business give lip service to protecting us, but often they do just the opposite, and do it knowingly.[2]

As the magnitude of all of this churned in my mind, I thought at first that maybe this book should be written by a person with a strong scientific background. It took me months to realize that what was needed wasn't a book about science at all but an investigative exposé about the politics and greed that fuels environmentally caused illnesses. It doesn't take a rocket scientist to see the terrifying patterns.

In fact, I was to learn that what the Bush administration calls the quest for "good science" is often just a diversionary tactic, a grand hoax, a fairy tale like "The Emperor's New Clothes," perpetrated by people who work together. The linguistic doublespeak allows the government, the Pentagon, and enormous corporations to mask urgent issues of public health and contamination so that they can continue to make huge sums of money even in the face of spreading epidemics and massive evidence of harm. Sometimes, the illnesses result from careless environmental contamination of our air, water, or food supply. At other times, toxic chemicals are deliberately thrust upon an unsuspecting population. Either way, the real criterion, the driving force, the bottom line that ties autism, asthma, cancer, birth defects, and cluster illnesses together is profit.

For me, as a mother and journalist, the heartbreaking stories of the individual victims I had come to know became the driving force. It was the women who were dead or dying of breast cancer or lung cancer, the children gasping for breath from asthma, suffering from autism, or dying of leukemia, and the babies who were born deformed or impaired who finally convinced me that I must write a highly accessible book for the general public, a book that would synthesize the issues and the expert opinions, a book that would spell out the moral issues, the spiritual issues, and the tragic human costs of remaining passive.

In some very real and threatening ways, government and industry have become life-and-death adversaries to the people who inhabit the earth. In the early 1990s, for example, the Department of Health and Human Services had requested, but not demanded, a recall of all vaccines containing thimerosal, a known and dangerous neurotoxin. But rather than offend the multibillion-dollar pharmaceutical companies, the FDA left it up to them to decide and allowed all fifty vaccines to remain on the market.

At the time of this writing, in the summer of 2006, that neurotoxin is still present in an unknown number of private and public health facilities. In fact, it is now being given to American children and adults, including pregnant women, in flu shots and some multidose shots, simply because it is less expensive to produce than the alternatives. It is also routinely being shipped overseas despite a great amount of evidence that it is creating epidemics of autism where none had previously existed. Efforts are also still being made to convince the public, especially American mothers, that there is still no "proof" of harm.

No one can adequately predict the long-term effects of thimerosal, let alone all the combined chemicals we are exposed to, and no one can control the idiosyncratic thresholds at which our individual bodies can no longer assimilate a particular load or combination of poisons.

Corporate scientists working in their laboratories can sometimes barely recognize what Albert Schweitzer once called "the devils of our own creation." Whenever dangerous evidence does begin to emerge, political leaders, big chemical companies, large corporations, drug companies, the military, and frequently even the EPA, the FDA, and other trusted watchdogs pit themselves against everyday people, denying responsibility and actively burying evidence. As a result, we, our children, and our children's unborn children are all being forced to assume the risks, without ever being told what they are or even that they exist.

As I read Rachel Carson's *Silent Spring*, I found that its warnings are even more relevant today than when the book was first published in 1962. Her primary concern had been with the endless stream of insecticides and pesticides we use to kill weeds, rodents, and other organisms we considered pests. She knew that these chemicals had the power to kill "the good with the bad" and asked, "Can anyone believe it is possible to lay down such a barrage of poisons on the surface of the earth without making it unfit for all life?"[3]

She also pointed out that "one chemical may act on another to alter its effect. Cancer may sometimes require the complementary action of two chemicals, one of which sensitizes the cell or tissue so that it may later, under the action of another prompting agent, develop a true malignancy."[4]

Rachel Carson struggled with breast cancer for four years and lived for eighteen months after *Silent Spring* was completed. During that year and a half, she was ridiculed by the chemical industry and awarded every possible prize from the world of journalism and letters. She was deeply relieved that she had lived long enough to complete *Silent Spring* but longed to go on to other important projects. Even after tumors in her cervical vertebrae caused her to lose function in her right hand, she refused to discuss her own illness, because she did not want to lose the appearance of scientific objectivity. Nevertheless, when she appeared at press conferences and spoke on national television and before Congress, Rachel Carson looked like a woman struggling with cancer; she wore a dark wig and suffered from the puffiness of radiation. Those who knew her noticed the dramatic change in her appearance.

Though Carson never mentioned her own cancer either publicly or in *Silent Spring,* she knew and wrote that many human cancers were directly linked to man-made chemicals and pesticides. Industry called her a hysterical woman who had overstepped her knowledge of science. They tried to malign her character and her research.

Today we live in a world gone far wilder than Rachel Carson's world, a world in which the most obvious toxic abuses are tolerated and global contamination has become a fact of life. Victims of cancer, birth defects, autism, and cluster illnesses must provide definite "proof" that a specific chemical or company has caused their problems before anything is done to protect them, and even then they are often paid large sums so that they will go away and die silently. The companies rarely admit guilt.

Not just America but the entire human population is now caught in the terrible conflict between our human inventiveness, brilliance, and creativity and our arrogance, greed, and shortsightedness in failing to care for the earth and its people.

Who is able to judge the specific load of toxic chemicals that each of our vulnerable unborn or growing children can tolerate? Who can tell us what will happen when two or more of these carcinogens act together or what is really a "safe dose" of PCE, TCE, perchlorate, mercury, thimerosal, or thousands of other poisons? Who can know your particular idiosyncratic sensitivities or previous exposures or those of your son or my daughter?

I have written this book because I do not want other people to experience what the people I met on my journey have experienced, and because I know that countless millions more will.

I come to this project with no special interest other than a concern for the future of humanity. I am not a politician, a religious leader, an environmental extremist, a scientist, or a board member of a large corporation. I am simply an investigative journalist, a woman, and a mother who has inadvertently fallen upon an urgent matter: the slow and not-so-slow poisoning of all our people for the short-term profit of a reckless few.

# 1

# GLOBAL CONTAMINATION

## No One Is Spared

> The entire cosmos is a cooperative. The sun, the moon, and the stars live together as a cooperative. The same is true for humans, animals, trees, and the earth. When we realize that the world is a mutual, interdependent enterprise . . . then we can build a noble environment. If our lives are not based on this truth, we shall perish.
>
> —BUDDHADASA BHIKKHU (1906–1993), A HIGHLY REVERED THAI MONK, OFTEN CALLED THE SERVANT OF BUDDHA, WHO LEFT BEHIND A LEGACY OF TEACHINGS

### PCBS, DDT, AND FLAME RETARDANTS

THERE IS A VILLAGE IN GREENLAND SO REMOTE THAT ITS 650 inhabitants still live almost as simply as their ancestors did four thousand years ago. It is one of the harshest, coldest, and most beautiful regions on earth. The native people of Qaanaaq, who call themselves Inuits, pride themselves on eating the food they harvest directly from the sea.

Many still travel in handmade kayaks and dog-pulled sleighs crafted from driftwood. Even today, their cultural identity is rooted in storytelling, dream dancing, and especially the passing on of ancestral techniques of ice hunting for food.

Looking at the way they live, at the magically glistening snow and the sparkling, clear blue Arctic sea, anyone would think that this was one of the purest, least tainted places on earth. They would think that chemical contamination from the big cities of the world couldn't possibly affect food or breast milk or children's health here. But as Marla Cone of the *Los Angeles Times* reported, they would be wrong—dead wrong.

For no obvious reason, about a decade ago a cluster of strange illnesses began to afflict the youngest children of the village. A large number of newborns unexpectedly began to show signs of serious neurological damage. Soon great numbers were becoming ill. These children were born with depleted white blood cells, their immune systems dangerously impaired and their brain development altered.

To some of the mothers it must have seemed like a dark spell from the gods. How else could they possibly have explained it? Surely they would never have thought that the ancient northbound winds and the deep ocean currents had carried toxic remnants from faraway lands directly into the food supply they hunted from the ocean. How could they have imagined that the trade winds carried more than two hundred dangerous chemical compounds that settled in the ocean and were poisoning their food and sickening their children? How could they have dreamed that the contamination started with the smallest ocean-dwelling creatures but slowly intensified as it traveled up the food chain into the larger fish and sea animals, then into the children and the adults who consumed them, and finally into the breast milk of the nursing mothers who passed it on to their vulnerable infants?

Although the children were the first to show the outward signs of poisoning, the bodies of the mothers actually contained the highest concentrations of polychlorinated biphenyls (PCBs) and other industrial chemicals and pesticides found anywhere in the world, levels so shockingly extreme that the breast milk and tissues of some Inuit women could actually be labeled hazardous waste.[1]

According to the Quebec Health Center, a concentration of

1,052 parts per billion of PCBs could be found in the breast milk of these mothers. The U.S. Environmental Protection Agency safety standard is just 2 parts per billion; 50 parts per billion is often considered to be hazardous waste.

The first hint of trouble came in 1987, when Dr. Eric Dewailly, an epidemiologist at Laval University in Quebec, began studying contamination in breast milk. Dewailly got a call from the lab's director saying that something was terribly wrong with the Arctic breast milk. The chemical contamination was off the chart. The peaks were overloading the lab's equipment and literally running off the page. The lab technician assumed that the samples of milk must have been tainted somehow in transit.

But they weren't. After checking more breast milk, Dewailly realized that the samples were accurate. The Arctic mothers' breast milk was seven times more contaminated with PCBs than the breast milk of nursing mothers in Canada's largest industrial cities and in other cities around the world where these man-made chemicals were used in the manufacture of electrical equipment and hundreds of other commercial products like paints, plastics, and dyes.

Still reeling from the data, Dewailly contacted the World Health Organization in Geneva, Switzerland. The expert in chemical safety with whom he spoke said that these were the highest levels of PCB contamination he had ever seen. He also said that the women should be told to stop breast-feeding their babies immediately.

When Dewailly hung up the phone, his personal conflict was intense. His head was spinning. He wanted to do the right thing, but how could he tell these isolated mothers that their breast milk was so contaminated they should stop breast-feeding their infants, especially when he knew that there was no formula available to feed them?

More than a generation has passed since that day, and Dewailly, who is one of the world's leading experts on contaminants, has continued to research the subject. He says that today it is not just PCBs that are poisoning the breast milk and sickening the children. DDT, the insecticide used to kill mosquitoes in tropical countries, and airborne flame-retardant chemicals that are usually found in electronic products and furniture have been transmitted through the atmosphere. They, too, have made their way into the food chain and are concentrated in fatty tissue. They build up over a lifetime, are stored inside the mother, then pass into the womb and contaminate the infant before he or she is even born. After birth, the baby gets another dose of poisons from breast milk each and every time he or she nurses.

"We have clusters of sick babies," Dewailly said. "They fill the waiting rooms of the clinics."

The average Inuit mother, like so many of us in America, Europe, and Asia, doesn't worry about the dangers of invisible chemicals. Instead, she continues to be concerned about the things she can see, taste, and touch, like how thin the ice is getting and where to get the family's next meal. Virtually every day, Inuit women continue to eat seal meat and whale meat and, with every bite, more PCBs, more DDT, more flame-retardant and other toxic chemicals continue to build up inside them. Then, with every pregnancy and every drop of breast milk, these mothers continue to contaminate their own vulnerable children.[2]

Wild animals in the region are also being contaminated. In 1998, several concerned Inuit hunters and elders began to collect specimens from animals that were exhibiting gross abnormalities. They included blind caribou, hairless seals, and polar bears born with male and female sex organs.

"We are the land and the land is us. When our land and animals are poisoned, so are we," says Sheila Watt-Cloutier, president of Canada's Inuit Circumpolar Conference, a group that defends the rights of Inuits. "We've always said that the Arctic is just the canary in the mine and that it's only a matter of time until everybody will be poisoned by the pollutants that are being created."[3]

Watt-Cloutier's concern is clear: "When women have to think twice about breast-feeding their babies, surely that must be a wake-up call to the world."[4]

# 2

# DEADLY WATER

## Cluster Illnesses, Leukemia, and Birth Defects

> For humans to injure other humans with disease . . . For humans to contaminate the earth's water, its land, its air, and its life with poisonous substances, these are sins . . . We have become uncreators. The earth is in jeopardy in our hands.
>
> —ECUMENICAL PATRIARCH BARTHOLOMEW, ARCHBISHOP OF CONSTANTINOPLE AND NEW ROME, CURRENT LEADER OF THE WORLD'S 300 MILLION GREEK ORTHODOX

### CHLORINATED HYDROCARBONS

"THE MILITARY STILL SAYS THEY DIDN'T KNOW BUT, BY THE LATE seventies, the signs at Camp Lejeune in Jacksonville, North Carolina, were everywhere. We had small animals coming out of the woods and dying all over the place. Birds would literally fall out of the air and flap a little bit on the ground and grow silent. Possums and raccoons and turtles were dying all around us. Every time we went outside, we could see more animals lying around on the ground," Ellen Harris explained, her voice husky with emotion.

"Some people thought maybe it was the mosquito spray. Others suspected that our Marines had picked something up overseas. Men, women, and children had clusters of strange symptoms and bizarre illnesses. We prayed about it and talked about it and cried over it, but we had no idea what it was. So, we just kept right on drinking the poisoned water and cooking with it and bathing in it.

"I was twenty-one and healthy when I came to Camp Lejeune in 1971. I'd never been sick with anything. I was just another young mother with two babies. Right now, at fifty-four, I have sores and tumors on every part of my body. My doctor says I need a double mastectomy. I'm in pain all the time, and my condition is terminal. My immune system is so damaged that I test HIV positive even though I'm not.

"Back then, I had an English bulldog that gave birth to seven puppies. Within a week, they had all died. First they went into muscle spasms; then they hemorrhaged from the mouth and rectum. After the second puppy died, I called the base vet. He seemed very concerned. He told me to wrap the puppy up and put him in a jar and said they would send him to the CDC in Atlanta. He sent a driver with a staff car and flew the puppy out of New River to the CDC. But whenever I contacted them at the CDC, they told me the study wasn't complete. Later, I found out that the study was complete, and the Marine Corps had actually received a report, which they buried, but every time I called they said they didn't know what I was talking about. They wouldn't tell me anything. No one ever gave me the results.

"At about that time, my youngest daughter suddenly began to grow hair all over her body, her arms, her legs, even her face. She was only two years old, but her hormones had gone wild. Her whole system was affected. She was sick all the time. She needed special education. She lost 50 to 70 percent of her hearing due to ear infections and had migraines, allergies, and asthma.

"Then my older daughter, Elizabeth, got sick. First, her temperature spiked. Then, she lost control of her arms, legs, and neck. I'll never forget her lying in the gurney in the emergency room. She looked up at me and said, 'It's all right, Momma; Jesus is coming after me.' It turned out that many other people had the same symptoms. No one knew what was going on. We were told that maybe contaminated spores were brought back on the clothing of a Marine who had returned from a Mediterranean cruise. They pointed to a sergeant major. But my little girl had never been near him. Nothing made sense. My daughter survived but was left with an immune system disorder that she still has today. Her first baby died before it was born. Even my surviving grandson, David, is chemically poisoned; we were told that it was passed on from his mother while he was still in the womb. I know this because we had him tested. He has allergies, asthma, and raw spots in his esophagus, his stomach, and his intestines.

"My husband was a Marine through and through, a man of honor, a gunner on a heli. The last week of his life he was in tremendous pain both from his war injuries and from the contamination. He had swelled up from gastric poisoning and turned a bluish or purplish gray. The week before he died, he made a deathbed confession. He told me that the military had been mixing chemicals to be used for war and then dumping them into the ditches. He had top-secret clearance and worked on chemical and biological development, but no one told him that those chemicals flowed directly into our drinking water supply. We were all deceived."

Ellen paused. "I grew up in the Marine Corps," she said. "And we were raised to love it. My father was in the Marines for thirty-one years. He was highly decorated and known as a China Marine. Every morning, I woke up at 4:00 A.M. to 'The Halls of Montezuma.' My dad always told me that anything you believe in is worth fighting for. I warned those in command that I would fight very hard to live so I could see this settled. We have a joke. It's a sick joke, but we say, 'Why send our troops overseas to fight our enemies? All we need to do is send them some of the water from Camp Lejeune. We could bottle it up and destroy nations.' We laugh, but it isn't funny. The tragedy is that they knew for years and they let us keep drinking it."

William C. Neil, who was assigned to test the drinking water at Camp Lejeune, had long since confirmed what Ellen Harris told me. In 1980, he had sent a warning of his own to those in command.

As *The Washington Post* reported, he wrote "'WATER HIGHLY CONTAMINATED WITH . . . CHLORINATED HYDROCARBONS!' in capital letters on one of his surveillance reports in early 1981. A private firm followed up with tests the following year. One of its samples showed an astonishing result: 1,400 parts per billion—280 times the level now considered safe for drinking water—of trichloroethylene (TCE), a likely cancer-causing chemical used for degreasing machinery that can impair the development of fetuses, weaken the immune system, and damage kidneys and liver. Other samples showed as little as 1 part per billion to 104 parts per billion—more than 20 times the level now considered safe—of tetrachloroethylene (PCE), a toxic dry-cleaning chemical that can seep into body fat and slowly release cancer-causing compounds."[1]

When the *Washington Post* journalists asked why the wells continued to remain open and supply drinking water for the next five years, the Corps said it was because there were no federal drinking water regulations at that time for either TCE, the metal degreaser, or PCE, the cleaning solution. The military said the cleaning solution had come from a small family-owned dry cleaner that is still open today adjacent to Camp Lejeune's main gate.

According to the same article, "Some water contamination experts believe the lack of enforceable regulatory standards for the chemicals would be a weak defense if the case ever made it into the courts. 'Even in those days, that would have constituted pretty close to a drinking water crisis,' said Richard Maas, director of the Environmental Studies Department at the University of North Carolina at Asheville."[2]

"'Crisis' is not a strong enough word," Ellen Harris said. "Many times the women were so sick that they were holding a frying pan to cook dinner for the family in one hand and a jar to throw up into in the other.

"If this had been a disaster that had happened because of a terrorist attack on the base, we could have accepted it. But this was neglect and abuse. This was an act by our own military that flagrantly said, 'Your lives and your children's lives have no meaning to us.'

"We were wives who loved our husbands and mothers who loved our children. We were a group of women who spit-shined our husbands' shoes and boots. We pressed their uniforms and polished their brass. We were a rambunctious bunch, our husbands' advocates in uniform. When they deployed, we did it all. We were the Sunday school teachers, the room mothers, and the nurses. We were the other half of the military men. We walked with them in their boots, so now, when we see that flag lowered to half-mast in memory of another one of us who has died from the contaminated water cover-up, it makes us terribly angry. No one seems to care about what has happened to us."

There was a pause; then Ellen's voice broke. "When I look at the children who have died and are still dying of leukemia, the children who go through the chemo and lose their hair, when I watch as their eyes weaken, I am moved by the way they keep wanting to live. I believe it's because they are the offspring of Marines who have always been such determined warriors, and I know that even after their deaths, the spirits of these children cannot be at peace, because their siblings are still dying."

In 2004, Camp Lejeune's water contamination case finally drew the attention of the EPA's criminal enforcement division. They dispatched investigators to gather information about the history of contamination at the base. As *The Washington Post* noted, the Agency for Toxic Substances and Disease Registry (ATSDR) thought that the Camp Lejeune wells may have actually been contaminated since the mid-1950s. If they were right, it would greatly expand the number of people who were victims and include veterans like retired Marine major Tom Townsend, whose wife was pregnant with their third child, Christopher, in 1966. Mr. Townsend returned home from duty in Vieques, Puerto Rico, just before his son, Christopher, died of a heart defect at the age of three months.

Now Major Townsend shares reports with Jerry Ensminger, a former master sergeant who was also based at Camp Lejeune but retired in 1994 after twenty-four and a half years in the Corps. Jerry is a slim, tough-talking, dark-haired man in his early fifties, whose eyes still fill with tears when he talks about Janey, the nine-year-old daughter he lost to acute lymphocytic leukemia in 1985.[3]

I first met Jerry on a clear, cold Saturday morning in April 2004. He had agreed to give me a tour of Camp Lejeune. Although it was just a few miles from my ocean house, I'd never been on the base, because they don't let anyone in without a special pass.

Jerry was already there waiting when I arrived at the parking lot of the Fisherman's Wharf restaurant. As soon as I had climbed into his old truck, so crammed with legal documents that there was barely room for my legs, he looked at me with a sad intensity in his face and explained that acute lymphocytic leukemia is the most common form of leukemia in children under fifteen and the most common kind of childhood cancer. Thirty-five hundred new cases are diagnosed in the United States each year. Few causative factors, except for exposure to high doses of radiation and contaminated water, have been linked to the disease.

Jerry frowned and said, "I've got a mission, Loretta. My little girl died in my arms and fifteen years later I found out that the people I had faithfully served for almost twenty-five years knew she was being poisoned by the water all along." His eyes widened. "I'm possessed, *possessed* by this thing." Even if he hadn't said it, I would have known.

"I've spent hundreds of hours on this case," Jerry said, pointing to the now infamous Tarawa Terrace housing area where he had lived. "You know what they did to save a buck? They came across the road from all these commercial septic systems, dry cleaners, gas stations, and auto repair facilities, and they drilled drinking water wells downgradient from them. They knew exactly what they were doing. But evidently, they didn't care. Even back in the 1950s, they knew about gradients and chemicals and the contamination of water."

"Do you know any of the people who live here now?" I asked.

"Hell no," Jerry replied. "I retired in '94, and I was a master sergeant then. This is all junior enlisted housing."

"I asked because I'm wondering how safe the water they are drinking here today is," I said.

"To tell you the truth, I'm worried about that, too. They only test once a year and they have a history of dishonesty. I'm also concerned about what's happening to the water on other military bases all across America," Jerry responded, taking his eyes off the road for a moment. "The water here now comes from six different wells. I'll show you some of them when we go to the main part of the base. At any given time, you've got forty-some thousand Marines and sailors living here, plus their families. This is a huge place, about two hundred square miles. It has its own hospital, its own capabilities for generating electricity, its own shopping center, high school, elementary school, and middle school. And, of course, it has its own water and sewer systems.

"This was the water we were swallowing every day, but they didn't tell anyone, and they didn't do anything. Finally, in January 1982 when the EPA and the Safe Drinking Water Act made it law, Camp Lejeune began to send water samples to Grainger Laboratories, which was a state-certified lab.

"It wasn't until two more years had passed, in June of 1984, when the EPA created a maximum allowable contamination level for VOCs [volatile organic compounds], that they got nervous. Five years. It took them five years," Jerry said as he slammed his hand on the steering wheel. "Five damn years and my daughter's life." He tossed his head to the left. "Why the hell did it take them five years?"

In the bitter silence that followed, we continued to drive.

"Here's what they call Piney Green Road," Jerry finally whispered, after he'd regained some of his composure. "This used to be dirt, but they paved it to take some of the pressure off the main gate at rush hours. See over there? That was a liquid waste disposal site." He pointed toward the woods. "First they dug the trenches; then they brought the toxic liquid waste out here and began dumping it into the trenches right beside the drinking water well. That's what Ellen Harris's husband confessed to just before he died. You see that fence back in the woods? There are signs on it saying, HAZARDOUS WASTE SITE—KEEP OUT."

I looked to the left and nodded.

"Well, that's what we were drinking. Can you imagine?" he asked. Then, without waiting for a response, he added, "First it was a drinking water well; now it's a 'cleanup site' for the huge toxic plume under the ground."

I must have looked puzzled.

"A plume is a column of fluid that flows underground as it moves away from its source. If it's toxic, it spreads the pollution," Jerry explained. "We not only drank from that poisoned plume and took baths in it; our kids made lemonade and Kool-Aid with it. They sold it to each other for five cents a glass and shot it out of water pistols.

"Look at those monitoring wells. That one is bubbling over right now. I guess the water table must be pretty high. With all the contaminated plumes they have on this base, they only tested once a year. Stupid, right?" Jerry's lips pressed together tightly for a moment. "No, wrong. It's because they didn't want to know. Finally, when they could no longer hide it, they tried to blame it all on the little mom-and-pop ABC dry cleaner across the street." Jerry shook his head for emphasis, then added, "They knew the dry cleaner's problem was just a drop in the bucket, but they used it to create a public affairs strategy that would focus all of the attention away from them. It was a typical military scheme."

As I listened to Jerry Ensminger, I thought about a warm, energetic, attractive woman I'd gotten to know at the church in Wilmington, North Carolina, that my son had been attending. Her family still owned and operated the ABC dry-cleaning business. She had talked to me about it before either of us knew I was going to write this book. She had explained that they were not allowed to sell the store because it had been designated one of the nation's worst toxic waste sites, a Superfund site. Then, one Sunday during a church coffee hour, she cried as she told me that her father had died of a heart attack during the investigation. Her mother now had cancer. The family had paid hundreds of thousands of dollars in fines. She said it had ruined them financially and torn them apart emotionally. Her father swore to her before he died that he had never known that the chemicals he used for dry-cleaning clothing could create contamination problems.

"Would you like to talk to the owners of the ABC dry cleaner?" I asked Jerry, knowing that in a legal sense they were adversaries yet feeling that they were all victims.

"Maybe someday," he answered evasively as he pulled into a parking place at the base hospital and pointed toward the main door. "I spent a lot of time here," he said, changing the subject. His voice grew husky. "Janey was diagnosed right in there."

Jerry's eyes drifted away from the hospital. "I got a call the other day from a man who is dying of cancer. He was on this base from 1963 to 1981. He said to me, 'Jerry, I'm dying, and there are some things I need to get off my chest.' When I went to see him, he told me the same thing Ellen Harris had heard from her dying husband. He said that the first day he started working at the base, they took him back and showed him the trenches at Piney Grove. As part of his job, he was told to dump the chemicals right into those trenches. Then, after that was outlawed, he and the other workers took the chemicals into the woods and buried them, not just a few gallons but thousands and thousands of drums of solvents that leaked directly into open pits, then seeped into the nearby drinking water wells. And do you know what he said they told him?"

I shrugged and waited for Jerry to speak again.

"They said, 'If you want to keep your job, you'd better keep your mouth shut.' I have it all on tape."

Jerry paused, then drummed his fingers on the steering wheel. "Sometimes, I still remember all those years of wanting to beat my head against a brick wall, because I couldn't get anyone to listen. Now things are finally happening, and those Marine commanders are scared to death. Hell," he added for emphasis, "it's their turn now. I joined the Marines at seventeen. It was my whole life, my whole world. I trained over two hundred young men and you know something? It makes me nauseous just to remember that now.

"The problem isn't just here at Camp Lejeune. They put a pretty face on these bases. They make them look real nice, but what's down under the surface is an environmental disaster, and I don't mean a disaster that's waiting to happen. Out of 168 federal contamination sites, the military already accounts for 130 of them, and do you know what they want now? They want permission to make it even worse. They want to get immunity from the Safe Drinking Water Act, the Clean Air Act, and the Endangered Species Act.

"I'm going to Washington to fight against those immunities. Congressman Dingell has asked me if I'd assist them with my testimony. He came to me because childhood cancers like my daughter's cancer have been directly linked to these chemicals. So I said, 'Congressman, you can count on me. I'll do anything I can to put an end to this.' I didn't start out this way. I was a registered Republican before I found out that most Republicans don't care about anything but big business and money."

After driving around the base and talking for nearly two and a half hours, Jerry pulled some candy bars and a Coke out of a brown paper bag and offered to share them. We were both hungry and tired of trying to talk over the engine noise of the truck. I asked if I could take him somewhere for lunch.

He shook his head. "Not on the base. I wouldn't want to talk about anything here. We could be easily overheard."

We headed back to Highway 17, back toward the Fisherman's Wharf in Jacksonville, North Carolina, where I'd parked my car.

The Fisherman's Wharf is probably one of the better restaurants along the tawdry strip of Route 17 largely populated by bars and "gentlemen's clubs." The restaurant was a large, older place that seemed to specialize in fried seafood, French fries, and hush puppies at upscale prices. It sat on the water, with a lovely deck that ran around the back for outside dining on warm days. It was too cold for that, but I was hoping for a spot that was quiet enough for my tape recorder to pick up Jerry's voice as he explained the legal and scientific documents he was carrying in his old leather attaché case. As it turned out, the place was crowded. There was another couple seated beside us and loud rock music competing with a noisy air conditioner that rumbled in the background.

"It was 1997 before I finally found out why my daughter died," Jerry said, speaking over the noise. "Even then, it was just by chance. A local TV station had picked up the story. The evening news was turned on in the living room, and I was carrying my dinner in on a plate from the kitchen. All of a sudden, I heard the newscaster saying that the water at Camp Lejeune had been highly contaminated from 1968 to 1985 and that the chemicals it contained had been linked to childhood leukemia. When I heard that, I just dropped my plate of food right on the floor and began shaking. The next day," Jerry added, breathing more quickly, "I started reading everything I could find and making contacts with everyone I knew.

"There are stages you go through when you lose a child to a catastrophic illness," he said. "First you go into shock; then you start wondering why it happened to your child. So, years ago, I checked my family history and her mother's, and found there was nothing on either side. But that nagging question of why Janey got leukemia had stayed with me throughout her illness, her death, and for fourteen and a half years after it. And, in that moment, that one moment, when I heard the newscast and dropped the plate of food right out of my hands, it all became clear. I suddenly knew why my little girl had died."

The restaurant had become quieter and the people at the next table had left. For a while, even the air conditioner cycled off and the music was playing old sixties tunes.

Jerry took out a photograph of a slender little dark-haired girl sitting on a bed. As he handed it to me, once again his eyes filled with tears. "This is Janey," he said.

"She was lovely," I responded, somewhat startled by her extraordinary delicacy and beauty. "What was she like?" I asked. "Can you tell me about her?"

Jerry stopped eating, and when he spoke, his voice had changed again. This time it had become so tender it was almost hushed.

"She was a daddy's girl," he said. "Fragile as she looked, she was a lot like me, the closest thing to a son I ever had. Whenever I'd go hunting, she always wanted to come, you know, just to hang out with me." He wiped his eyes. "I signed up with the Marines to serve my country, but I never signed anything that gave them the right to kill my child, to knowingly poison her."

I must have looked skeptical.

"No, I mean it." His voice rose. "I've said that publicly, and the Marine Corps has never refuted anything I've said, including that the contamination went back to the 1950s. I'm sure they know that from the geological studies. Sometimes people ask me if I'm afraid they will try to hurt me for stirring things up this way, but, you know what, I'm not afraid of anything. The connection between leukemia and contaminated water has been confirmed and I'm leading this fight for everyone. Besides," Jerry added with what sounded like a forced laugh, "can you imagine how obvious it would be if anything suddenly happened to me now?"

For a moment my mind jumped to the rumors surrounding Karen Silkwood's death and the way she was allegedly pushed off the road as she drove to meet a *New York Times* reporter carrying evidence about plutonium contamination at the Kerr-McGee plant, evidence that was never found. I looked at Jerry from across the table and noticed that he had pushed his plate aside even though it was still half-full.

"The day she died . . . ," I began haltingly.

"The day she died," Jerry repeated, slowly and with an intensity that stopped me in midsentence. "Janey was in a lot of pain, so they suggested that she take morphine. She didn't want to, because she had tried it before and it made her so tired. But this time she just couldn't handle the pain.

"Janey went through hell for nearly two and a half years, and I went through hell with her," he continued, clearly wanting me to know the whole story. "I was there with her every step of the way. Her mother couldn't handle it. Every time Janey went to the hospital, I was the parent who went with her. I slept in the room with her. It was always me holding her. Sometimes, she was screaming in my ear, 'Daddy, don't let them hurt me.' Like when she had the bone marrow transplant and the spinal taps. The last time she went into the hospital was the last day of July, just before her ninth birthday. She didn't come out until September 20, and that was in a casket.

"About a week before she died, the head of Hematology had come in talking about a new form of therapy. He said it would cause severe burns and ulcers, and they didn't recommend it, but Janey looked at them, blinking to control her tears, and said, 'This is my life you're talking about, and I'm not giving up. Let's try it.'

"The ulcers were all over her mouth, her legs, inside her nose and her vagina. The day she died she was in such intense pain she could hardly speak, but finally she managed to whisper, 'I want to die peacefully.' When Janey said that, I started sobbing. She hugged me and said, 'Stop it. Stop crying.'

"I said, 'I can't help it. I love you.'

" 'I know you do,' she replied. 'I love you, too. But, Daddy, I hurt so bad.'

"Do you want some morphine?' I asked. She was already being given methadone.

" 'Yes, Daddy' she said. 'I'm ready.'

"When the nurse heard that Janey wanted morphine, she knew the time had come. Then, just as they started to give it to her, she said, 'Wait. Stop. I want some for my daddy, too.'

" 'This is a very powerful pain medicine. I can't give it to your daddy,' the nurse said.

" 'But my daddy hurts, too,' Janey answered."

Unashamed, Jerry wiped his eyes with the back of his hand. "You see, I always took a little of whatever she took to show her I was with her," he managed. Then he shrugged his shoulders and closed his eyes for a moment. "The morphine killed her.

"It's a respiratory depressant," he added. "As soon as she died, a man rushed in to resuscitate her, and I jumped up and put my hands on his shoulders and said, 'If you touch her, I'm going to kill you.'

"And then to find out that the organization I served faithfully for twenty-four and a half years knew about this all along, and they never said anything, well, shame on them. And God help anybody who gets in my way."

The next time I saw Jerry Ensminger was April 19, 2004. We were both attending a public hearing in Jacksonville, North Carolina. I was there not only because I knew Jerry was planning to speak publicly about Camp Lejeune's contaminated water but also because I had received a notice with my own water bill saying that plans were moving forward for several communities that were running short on fresh water to lease drinking water and sewer disposal from Camp Lejeune. It seemed the base had surplus water that, for a price, would serve the needs of several communities, including my own. In fact, once the arrangement was approved, their drinking water would begin supplying my house on the ocean.

Hundreds of home owners piled into the old courthouse that night. They filled every seat in the place, then stood on the sidelines.

The meeting was long and contentious. Onslow County's public utilities director, Kerry Randel, stood up and said, "Until and unless Lejeune cleans up its act, you would be ill advised to take that water."

When it was Jerry's turn, he stood behind the lectern and acknowledged me with his eyes. Then, looking both determined and pained, he read from a prepared statement.

"I am here this evening to inform the citizens of Onslow County of the facts about contamination plumes that still exist in the groundwater aquifers underneath Camp Lejeune. But before I do that, let me give you some of the history regarding past contaminated drinking water on base and how the military dealt with it."

His passion was palpable to everyone as he told Janey's story, then explained that beginning in 1999, the ATSDR had conducted an adverse pregnancy outcome survey for women who were pregnant and living on base between the years of 1968 and 1985. In that survey, they had identified thirty-three neural tube defects, forty-one oral clefts, seven childhood lymphomas, and twenty-two childhood leukemias, including Janey's. There were 103 sick or deformed children among the 12,598 who had been exposed to Camp Lejeune's contaminated drinking water before they were born. Jerry said that the twenty-two cases of childhood leukemia identified in the study was about sixteen times higher than would be expected under normal circumstances. He explained that the figure was based on average cancer rates among control groups in nine other studies looking at children born between 1975 and 1978, then added that no survey or study had ever been conducted or proposed for the children who became ill while living at Camp Lejeune if they were not born there.

I was sitting in the front row, and I could see his hands shaking as he spoke. I knew that he was referring to about 8,000 more children who had purposely been excluded from the study.

"On September 24, 1985, Janey succumbed to her disease," Jerry added, looking steadily at the audience. "She was only nine years old. It has been suggested by researchers that as many as 90 percent of all childhood cancers are suspected to have been caused by environmental exposures. My question to all of you this evening is an ominous one. What lies in the future for the children of Onslow County and the families that love them?"

The ATSDR has established that multiple sources of chemical contamination still exist in the groundwater at Camp Lejeune, and "it is unknown whether contaminant levels are increasing or decreasing."[4] Nevertheless, the commissioners voted three to two to lease Camp Lejeune's water and send it out into the civilian community. They also agreed to a thirty-year water lease with Camp Lejeune and to an automatic fifteen-year extension.

Worse than the shock of learning about the tragedies at Camp Lejeune and realizing that now the people in the community where I lived would be drinking and bathing in Camp Lejeune's water was my growing awareness that leukemia caused by environmental chemical poisoning would shape the future of children all over the world, especially children living on or near military bases.

According to *USA Today,* the U.S. Department of Defense is the largest polluter in the world, producing more hazardous waste than the five largest U.S. chemical companies combined.[5]

In the "United States, one in every ten Americans lives within ten miles of a military site that has been listed as a Superfund priority cleanup site." As of June 2006, the EPA was refusing to release information on how much money and time it would take to clean up these 140 sites. That is because the Superfund has run out of money and been largely inactive since 2002. The Fund was allowed to run dry when Congress failed to renew the tax on polluters. About 11 million Americans, including 3 to 4 million children, live within one mile of these military sites and suffer increased risk of leukemia and other illnesses.[6]

In 2001, Sierra Vista, Arizona, and Fallon, Nevada, two small towns with military bases in close proximity, were simultaneously struggling with leukemia outbreaks.

Arizona officials had recorded six cases of leukemia in children up to the age of fourteen since 1995. Nevada officials had recorded fourteen cases of leukemia in children up to the age of nineteen since 1997. One patient, six-year-old Anastacia Warnecke, had lived in both towns.

Anastacia grew up in Sierra Vista, Arizona, and was diagnosed with leukemia in March 2000. She was counted among the Fallon, Nevada, cluster even though she had only lived in Fallon for two weeks. Because of that decision, Sierra Vista was able to deny that it had a cluster.

The Arizona Department of Health Services (ADHS) said, "It is true that one more case would put us into a statistically significant cancer cluster, and she would be that case. That is if she was included in the count." Instead, they decided that the girl would be considered an "associated case." They also admitted that health officials weren't anxious to begin another leukemia cluster probe in Sierra Vista, Arizona.

The Centers for Disease Control and Prevention had already investigated 108 other cancer clusters, including one in Arizona that went on for ten years, and had failed to find an environmental reason for any of them. There were two possible explanations for this run of negative outcomes. Either there were no links or the CDC didn't want to find them.

While Arizona continued to deny that they had a cluster, officials were about to begin an investigation in Fallon, Nevada, where state health officials had admitted that the rate of acute lymphocytic leukemia among children in the town was roughly 100 times greater than in the rest of Nevada.

"As a parent, I can think of nothing more heartbreaking than watching a child battle a life-threatening illness and nothing more frustrating than not knowing the cause of that disease," said U.S. Senator Harry Reid (D-NV), ranking Democrat on the Committee on Environment and Public Works hearing held in April of 2001. "When cancer strikes a community, we have an obligation to focus every available resource on finding and eliminating that danger. This hearing has brought together medical experts, state officials, federal agencies, and the community, all in hopes of gaining new insight into this mysterious cluster of leukemia cases and others like it around the nation."

Joining Reid at the hearing was U.S. Senator Hillary Rodham Clinton (D-NY), who had also served as a member of the committee. In addition to sharing Reid's concern with learning more about leukemia clusters, Clinton also wanted to identify constructive ways to improve the federal role in addressing all of the nation's environmental illnesses.

"I agree with Senator Reid," she said. "We need to do a better job of tracking environmental health effects, whether it is asthma, cancer, birth defects, or other disorders associated with environmental exposures. By doing so, we can work constructively, not just to identify and address these chronic disease outbreaks, but to prevent them in the future."[7]

Unfortunately, despite the good intentions of Reid and Clinton, it wasn't going to happen.

Adam Jernee was the seventeenth child from Fallon to die from acute lymphocytic leukemia, the same remorselessly fast-moving cancer of the blood that had killed Jerry Ensminger's little girl, Janey. Adam was only eight years old and had fought the cancer for more than two years of his short life. By 2001, it was also clear that leukemia wasn't the only illness with which children and adults in Fallon were struggling. They were also afflicted by a myriad of rare diseases that included myelodysplastic syndromes and aplastic anemia, diseases that relentlessly attack the bone marrow. According to the Leukemia Research Fund, "Myelodysplastic syndromes (MDS) are a group of diseases that are sometimes called preleukemia, in which the production of blood cells by the bone marrow is disrupted."[8] In leukemia, white cells are produced in excessive numbers. In MDS, the production of any and sometimes all cells is affected. The only viable treatment option is donor stem cell transplant in young and fit patients.

Aplastic anemia or bone marrow failure is "a rare disease that develops when the bone marrow fails to produce blood cells. The only real 'cure' for aplastic anemia is bone marrow transplant."[9]

Parents in Fallon were convinced that all of these illnesses were being caused by exposure to toxic chemicals. They were also certain that those chemicals originated somewhere on the 240,000-acre Fallon Air Station. According to *CounterPunch*,[10] a local political newsletter, "The Navy and the Pentagon have done next to nothing except deny culpability and try to bully anyone who demands answers from Naval brass." *CounterPunch* asserted that "the Navy has known about high levels of cancer among the children of Fallon workers and Navy officers since at least 1991."

*CounterPunch* quoted Brenda Gross, whose six-year-old son had been sick with leukemia for two years: "Our frustration level is very high. This should have been found out and stopped a long time ago. But you can't get anything out of the Navy."

Local residents thought the problem was jet fuel spills and fuel dumping by Navy aircraft. That's because the Fallon Air Base had at least sixteen toxic waste sites that were contaminated with JP-8 jet fuel. That fuel, "a combination of kerosene and benzene, is a known carcinogen that had previously been linked to leukemia and other bone marrow disease."

Records from the state of Nevada showed that several distinct plumes of the fuel entered the groundwater beneath the air base. According to same source, "Nearby residents charge that Navy fighter pilots routinely dump excess fuel into the desert prior to landing at Fallon." The Navy claimed this only happened about three times a year.

The jet fuel spills could easily be causing the cancers. However, according to *CounterPunch* editors, there could be an additional explanation. A study of area groundwater by the U.S. Geological Survey in 1994 clearly showed that at least thirty-one drinking water wells were contaminated with high concentrations of radioactive minerals. That contamination occurred when the U.S. Department of Energy detonated a nuclear device below the ground southeast of Fallon. Radionuclides that were released during the detonation had migrated into the water supply. The information was kept a secret for seven years while people continued to drink the water. Finally, in September 2001, a former U.S. Geological Survey (USGS) staffer came forward.

Despite all of the known contamination, the people of Fallon got no answers from the government's cluster investigation. "I think there's a potential cover-up here," said Richard Jernee, Adam's father. "I don't have faith in any of these people. How many kids have to die before we get the truth?"

"When are these people going to do something real?" asked Floyd Sands, whose twenty-one-year-old daughter, Stephanie, had recently died of leukemia. "I haven't seen them do anything real so far."[11]

In fact, the only things the federal cluster investigation turned up were a high amount of tungsten, a heavy metal used to make jet engines and nose cones for bombs and rockets, and a large amount of arsenic. Government scientists said they did not think that either of these was linked to cancer.

They claimed, however, that the elevated levels of tungsten and the arsenic in tap water were "very important findings" that would form the basis for more research in the future.

"We've learned a lot, but we haven't found the cause of this leukemia cluster," announced Dr. Carol Rubin, head of the research team from the Centers for Disease Control and Prevention in Atlanta.

They then asked the National Institutes of Health to begin researching possible links between tungsten and cancer.[12]

Deborah Frisch, a retired program director for the National Science Foundation and an expert in risk analysis, said, "The tungsten study is junk science. . . . It doesn't look like the government is interested in finding an answer."

According to the *Tucson Weekly*,[13] government agencies and studies do all sorts of things to avoid finding the environmental causes of cancer. They throw out cases; they limit samples; they accept industry testimony without question; they use extremely complex statistical methods and write conclusions that don't always match their data.

"A final report issued by the ATSDR went even further. Despite all the leukemia and other illnesses, it concluded with astounding certainty, 'There are no past, current, or future public health hazards from exposures to the Fallon Naval Air Station substances in the environment.' It's difficult for most people to understand how they could so confidently predict the future."

The *Tucson Weekly* asserted that two factors contributed to the government's cluster illnesses investigation: one was money; the other was politics.[14]

Although no cause of the leukemia cluster in Fallon, Nevada, was ever found, "a cluster of this magnitude would be expected to occur in the United States by chance only about once every 22,000 years."[15]

Meanwhile, in Sierra Vista, Arizona, federal authorities were still denying that they had a leukemia cluster and were still trying to avoid an investigation.

Three more cases were diagnosed in Sierra Vista in 2001. One additional case was diagnosed in 2002, but the family had moved away before the leukemia was identified, so the state cancer registry refused to include it. "Then, in 2003, a group of three more children were diagnosed, two of them within a month. This small town of 40,000 now had more than three times the number of leukemia cases doctors expected to find."[16] The cluster was finally undeniable.

To the parents of these leukemia victims, statistical arguments about whether the cluster is real or not are irrelevant and infuriating. As the *Tucson Weekly* reported, they have watched their children vomit constantly, lose their hair, and submit to blood tests and marrow biopsies. These parents ask about causes and get explanations of random distribution. When they continue to wonder why their children got sick or died, they're told to put the whole thing behind them.

Another disturbing cluster investigation had taken place in Maryvale, Arizona, an industrial neighborhood in West Phoenix that had seen forty-nine childhood leukemia cases beginning in

1965. For much of the duration of the cluster, there was no treatment for leukemia. Virtually every child diagnosed died. The investigation was conducted by the Arizona Department of Health Services (ADHS) under the oversight of the CDC. The media discovered that in 1982 the ADHS had told the parochial school principal who first noticed that children were dying not to talk about it.

This Arizona study cost millions of dollars and went on for more than ten years. It, too, found no link between the leukemia in the area and any environmental factors. The negative result surprised many people, especially since the city of Phoenix had already begun closing the neighborhood's drinking water wells, because they were dangerously contaminated. By 1987, the whole area had been identified as a Superfund cleanup site, but the ADHS study still refused to look at the water supply. Instead, they collected data on a multitude of other variables, including diet and TV ownership.[17]

Even if the Arizona studies had identified the contaminated water supply, it might already have been too late to contain it, because as Rachel Carson warned in the 1960s, it is not possible to add dangerous chemicals to water in one place without threatening the purity of water in other places. Since water keeps moving, it can seldom be contained in compartments that remain separate. Even rainfall eventually becomes groundwater when it passes down into the earth into the soil and rock. The water may rise up under hills and sink down beneath valleys, but ultimately, one way or another, contaminated groundwater just keeps moving. Except for what enters as rain or surface water, all running water was once groundwater, so the contamination of groundwater means the contamination of water everywhere.[18]

## PERCHLORATE IN WATER

Water is the basic source of all life and a primary symbol in religious traditions. It cleanses, purifies, refreshes, and inspires. The Bible speaks of a fountain of living water and of justice flowing as a mighty river. Without water, everything dies. It is the basic element through which all life-forms emerged, exist, and flourish. It has been the lifeblood of the planet for more than 4 billion years. Yet today water is threatened almost everywhere on earth.[19]

Perhaps the most disheartening of my own realizations in the fall of 2004 dealt with groundwater contaminated by perchlorate, a toxic chemical widely used in the manufacture of rocket fuel and explosives. Perchlorate, which has now been linked to thyroid cancer, Graves' disease, neurological disorders, devastating birth defects, and a long list of other possible conditions related to the thyroid, was originally believed to be harmless. Later, people were told that the toxic substance might be dangerous but would only affect water supplies near areas where the rocket fuel was actually made or used. In fact, perchlorate had soundlessly crept underground from the small streams, wells, holding ponds, and rivers into which it had often been carelessly dumped into major waterways that now supplied drinking water to countless millions of American families all across the country.

By 2004, perchlorate had been discovered in the drinking water of thirty-five American states. According to the EPA, those states were Alabama, Arizona, Arkansas, California, Colorado, Florida, Georgia, Illinois, Indiana, Iowa, Kansas, Maryland, Massachusetts, Michigan, Minnesota, Mississippi, Missouri, Nebraska, Nevada, New Jersey, New Mexico, New York, North Carolina, Ohio, Oklahoma, Oregon, Pennsylvania, South Carolina, South Dakota, Tennessee, Texas, Utah, Virginia, Washington, West Virginia, and the territory of Puerto Rico.[20]

In June of 2004, Senator Dianne Feinstein told a gathering of senators that "perchlorate contamination posed a major threat to the health of all Americans and that it was imperative that the Department of Defense (DOD) disclose all evidence of perchlorate contamination at DOD sites and begin cleaning them up as soon as possible."[21]

The Senate approved Feinstein's amendment to address and control perchlorate contamination. On June 22, 2004, they approved $4 million to clean up the pollution, but it meant little, because perchlorate was already completely uncontrollable. The chemical had spread quickly across the United States, especially in the Southwest, polluting, among other places, the lower Colorado River, which is now the water source for more than 20 million people. Most of those people live in Los Angeles, San Diego, and other Southern California cities that take water from the Colorado River.[22]

By 2005, perchlorate from Los Alamos National Laboratory (LANL) had reportedly reached the Rio Grande through underground springs. Once again, it was really no surprise, since the Los Alamos lab had been discharging contaminated water from its radioactive treatment plant since 1963. The Rio Grande supplies water to an additional 4 million people in Albuquerque and Santa Fe.

On May 23, 2006, a group of six New Mexico community groups filed a Notice of Intent to Sue the U.S. Department of Energy and the Regents of the University of California. " 'More than sixty years of contamination . . . now threatens our future drinking water supply,' said Kathy Sanchez, director of Tewa Women United, one of the community groups. 'There are now more than 1,400 documented contaminant sites at LANL, and every time it rains or snows, these contaminants move though our canyons and springs to the Rio Grande.'

"Matthew Bishop, of the Western Environmental Law Center and legal counsel for LANL Water Watch, said that the impending lawsuit was based on four specific violations of the Clean Water

Act: failure to conduct adequate monitoring, failure to report violations, failure to have pollution controls in place, and making unauthorized discharges. 'The result of these failures is that toxic contaminants are migrating to the Rio Grande, the future source of drinking water for Albuquerque and Santa Fe,' he said.

"Bishop called the pollution a catastrophe waiting to happen. 'In addition, the Rio Grande continues to be used for fishing and farming all along its length, enabling dangerous contaminants to get directly into the food chain.' "[23]

It wasn't until the 1970s, after the passage of the Safe Drinking Water Act, that the defense industry was first forced to try to control or contain perchlorate.

According to *The Wall Street Journal,* it took another fifteen years before perchlorate finally became publicly known as a dangerous threat to public health. Even then, in 1985, almost forty-five years after it was first used, it was only because "the EPA detected it in wells serving about 42,000 households near Aerojet's original facility in the San Gabriel Valley near Los Angeles. The agency found concentrations ranging from 110 parts per billion to 2,600 parts per billion."[24]

When the news of massive perchlorate contamination first became public, more than eight hundred people in Redland, California, about sixty miles from Los Angeles, began waging a legal class action battle for damages against Lockheed Martin.

In March 2003, the California Supreme Court blocked the suit, saying class actions were only allowed when all of those suing had similar claims. The justices said the only way to determine if a plaintiff had been injured was for that plaintiff to sue on his or her own.

Adrienne Wise-Tates was one of the people who did that. According to an article posted online at New West network,[25] she was a child when Lockheed Martin began using perchlorate. "When Lockheed ceased operations at the location in 1974, Wise-Tates was twenty-two." Her case was thrown out because the statute of limitation on reporting her injuries had passed.

Wise-Tates is quoted: "It's just that we didn't know anything about this until we were grown. When I started having all of these health problems, a friend said to me, 'Do you think it's because of the water?' Then I was in my early twenties. I had to do some investigation. I guess they claimed that I should have known all along, but I didn't know."

According to the same article, "Over the years, Wise-Tates has suffered a string of ailments, including 'thyroid problems, surgery on a goiter, cancer on my throat, open brain surgery, and three breast cysts.' She believes that all of this was a result of drinking water tainted by Lockheed.

"I played in the water, drank the water. All the normal things a child does," she said. "Since so much perchlorate was in the water, that's what I attribute it to."

Lockheed spokeswoman Gail Rymer said the company "vigorously defended" itself against the claims.

"We do not feel that anyone was harmed or has been made ill as a result of our operations at the former Lockheed Propulsion Company site," she said.[26]

The strategic tactic that allowed perchlorate to spread unchecked into our waterways for so many years was a so-called scientific debate about "safe limits" that took place between the EPA, the Pentagon, and its allies in the defense industry.

According to environmentalists, the strategy was simple. Although none of the players denied that the chemical was toxic, they all knew that as long as they fought over just how much perchlorate a human being could tolerate before becoming ill, the chemical would remain an unregulated contaminant and nothing would be done to control it or diminish the huge profits it reaped. As a result, in 2005 the United States was still years away from establishing a nationally enforceable "safe" standard.

The EPA wanted water supplies tested for perchlorate nationwide, but the Pentagon asked Congress for an exemption from all responsibility, claiming that "environmental regulations were a threat to national security since they restricted the military."[27]

During the arguments over safe perchlorate levels and the rising evidence of severe harm, eight states passed their own widely divergent advisory limits, ranging from 1 part per billion in Maryland, Massachusetts, and New Mexico to 18 parts per billion in Nevada.

The EPA had already warned that even the smallest traces of perchlorate were dangerous, especially to infants, who were at particular risk for neurological damage.

"After everything I've seen on perchlorate, I'm a lot more concerned about even subtle deficiencies of thyroid hormone on brain development than I was before,"[28] says biologist R. Thomas Zoeller, an endocrine expert at the University of Massachusetts at Amherst and one of the seventeen peer reviewers of the EPA's report.

In January of 2002, the EPA proposed a draft "reference dose" for perchlorate in drinking water of 1 part per billion. "The Pentagon and several of its major contractors, all facing billions of dollars in possible cleanup and liability costs, say that perchlorate should be allowed in drinking water in concentrations up to 200 parts per billion. 'The scientific basis for believing there's harm has not been established,' says Maureen Koetz, assistant undersecretary of defense for the environment."[29]

Everyone in the Defense Department knew that reversing the environmental damage, if it could be accomplished at all, would take decades and cost billions of dollars. That was why they suggested 200 parts per billion, then mounted a major campaign of delay, deception, and incorrect information designed to convince the press and the public that the dangers of perchlorate remained stuck in an unsupportable maze of scientific uncertainty.

By 2004, according to Dianne Feinstein's report to the Senate, "Perchlorate had not only been found in the Colorado River, but contamination had also been identified in at least 300 groundwater wells, and the list was still growing. Contaminated wells were found in eastern Sacramento County, Simi Valley, San Gabriel Valley, the Rialto-Colton Basin, and water sources for the Santa Clara Valley Water District."[30]

Attempts at remediation continued to be slowed by lawsuits and requests for federal immunity. Even California's effort to establish the nation's first standard for a "safe" level of toxic perchlorate at somewhere between 2 and 6 parts per billion was delayed because Lockheed Martin and Kerr-McGee forced California to submit its recommendations to outside review by industry-picked experts. As *The Wall Street Journal* discovered in their three-month study, Kerr-McGee had its own concerns. After perchlorate was found in drinking water taps in Los Angeles, "scientists traced the plume four hundred miles up the Colorado River to Lake Mead" and, finally, "to Kerr-McGee's giant ammonium perchlorate plant in Henderson, Nevada. The Navy built the plant in the 1940s to make perchlorate compounds for the war." After the war ended, the plant was "inherited by Kerr-McGee in a 1967 merger." During the next decade, they "spilled thousands of pounds of perchlorate waste . . . into unlined evaporation ponds. The chemical leached into shallow groundwater, seeping into . . . Lake Mead, the main drain . . . for wastewater coming from Las Vegas."[31]

According to Nevada's Division of Environmental Protection, perchlorate was first detected in Kerr-McGee's groundwater in the mid-1980s, but, predictably, it was ignored. At that time, the company was still injecting high levels of perchlorate back into the ground each day. "The guidance on perchlorate was lacking," Patrick Corbett, director of environmental affairs for Kerr-McGee, later claimed.

By 2004, Kerr-McGee was clearly in deep trouble. They had

spent roughly $70 million in an attempt to remove the perchlorate they had put into the water supply, but according to the EPA, they were "catching only about half the 900 pounds a day seeping into the Las Vegas Wash." They had also filed a lawsuit seeking reimbursement from the Pentagon for the cleanup costs and said they were "adding new systems to capture much more of the perchlorate." Despite their massive remediation efforts, "so much perchlorate had already entered Lake Mead that the levels below Hoover Dam—all the way out to Los Angeles—have hardly budged in five years."[32]

At the end of 2004, the EPA was still continuing to hold out against pressure for a level of perchlorate restricted to just 1 part per billion and the Pentagon was still insisting that 200 parts per billion were safe and acceptable.

Then, in January 2005, the highly respected National Academy of Sciences came out with a surprising report. They shocked the scientific community and undermined the EPA's suggested perchlorate level of 1 part per billion by saying that "safe" levels of perchlorate were about twenty-three times higher than the EPA had declared.

The Natural Resources Defense Council (NRDC), a national nonprofit organization of scientists, lawyers, and environmental specialists founded in 1970 and dedicated to protecting public health and the environment, decided to find out what was behind the new report. They sent more than a dozen Freedom of Information Act requests to the EPA, the Department of Defense, and the White House. All of those agencies "stonewalled them for more than a year." Finally, after suing, the NRDC succeeded in obtaining about thirty boxes of secret information. Even then, the White House and other agencies continued to withhold or black out thousands of additional documents or sections of documents that dealt with perchlorate contamination. A single-spaced list of the withheld documents, the "Vaughn Index" that the government finally submitted to the court, was more than 1,500 pages long.

Ultimately, after carefully studying the secret files they had obtained, the NRDC announced that the academy's recommendation on safe levels of perchlorate was the direct result of a covert campaign by the White House, the Pentagon, and defense contractors. The NRDC said that the academy had been "strong-armed" into making their perchlorate recommendation.[33]

The academy denied that it had been pressured. "The government had no influence over the conduct or outcome of this study," said E. William Colglazier, the academy's executive officer. "The committee members were highly competent, there were no conflicts of interest, and we have full confidence in the report."[34]

Despite the denials, the NRDC insisted that there had been "extensive involvement by White House and Pentagon officials to limit the scope of NAS's inquiry and select the panelists, as well as collaboration among the White House, Pentagon, and DOD contractors to influence the panel."[35]

The reason was money. If federal and state regulators accepted the academy's recommendation of perchlorate levels that were twenty-three times higher than the EPA's, the Defense Department could avoid cleanup costs of hundreds of millions of dollars.

The Defense Department had been blocking EPA efforts to deal with perchlorate for more than a decade. The pressure finally became so intense that the EPA deleted from their Web site its statement that 1 part per billion was the perchlorate level they considered safe. In 2004, the original version was still available at NRDC, but the "cleansed" version was posted on the EPA's Web site. In February 2005, the EPA quietly reversed its own position and adopted the National Academy of Sciences recommendation. They created a new perchlorate level of 24 parts per billion "without any public comment or review."[36]

Bill Walker, the West Coast vice president for the Environmental Working Group, a highly respected watchdog agency that has been at the forefront of conducting several groundbreaking scientific environmental investigations, warned, "Perchlorate interferes with thyroid function. Low levels of thyroid hormones can lead to birth defects and there's evidence that we are talking very low levels of perchlorate."[37]

According to same article, "An Arizona study showed that women who drank Colorado River water with perchlorate levels of 1 or 2 parts per billion had infants with different thyroid hormone levels from infants whose mothers drank uncontaminated water."

That is because perchlorate interferes with the thyroid's ability to make hormones that are critical to brain development. Unlike adults, infants can't store a supply. Thyroid hormones also play an important role in the development of other organs. Earlier studies had shown that even small disruptions in thyroid hormones in utero or during early development could lower a child's IQ while larger disruptions could cause mental retardation, loss of hearing and speech, deficits in motor skills, and birth defects.

Under pressure from all sides, California's government decided to issue its own preliminary threshold for perchlorate. As a compromise, they suggested 6 parts per billion, a number six times higher than the 1 part per billion the EPA originally said was safe but lower than the academy's recommendation of 24 parts per billion and much lower than the 200 parts per billion that the Pentagon wanted.

One person who was deeply concerned about California's compromise with perchlorate was Larry Ladd, a medical geologist whose own town of Rancho Cordova, California, had already lost twenty municipal drinking water wells to perchlorate contamination caused by Aerojet.

Since 1997, Larry had served as a full-time volunteer to the healing and assessment team for the California Department of Health Services and an EPA-sponsored group that performed a similar function.

"As soon as I found out what was going on in Rancho Cordova," Larry explained, "I began distributing leaflets to homes in local neighborhoods. I was especially worried about the perchlorate levels in houses near our local United Methodist church, where I'd heard they had gone as high as 250 to 400 parts per billion. At first, all I did was put leaflets on doors, warning people about the situation. Later, I decided to knock on some of those doors. Once I got started, I knocked on one door after another and kept finding thyroid-related illnesses and young women who were losing babies. Most were in their first trimester.

"Now, as far as we know, most of our contaminated wells in Rancho Cordova have finally been closed, but not before massive damage was done."

Larry was giving me background information as we drove from his home in Rancho Cordova to Morgan Hill, about one hundred miles south, where perchlorate contamination from the Olin Corporation had recently become a major public health concern. We were going there to meet Belinda Reanda at the Lyons Restaurant on Highway 82, just outside of town. She had lived in Morgan Hill since 1988 and had been diagnosed with Graves' disease, an autoimmune thyroid disease characterized by fatigue, weight loss, an enlargement of the thyroid gland, and other symptoms. That same year, her daughter, who was six years old, was diagnosed with Hashimoto's thyroiditis, another type of autoimmune disease in which the immune system attacks and destroys the thyroid gland. The child had developed extreme fatigue and had lost her appetite two years earlier but was continually misdiagnosed. Belinda, whose Graves' disease was being treated with radiation, developed such extreme chronic fatigue that she could barely care for her ailing daughter. "Perchlorate," Belinda explained, "started the downward spiral that all but destroyed my family's life."

Now, after years of multiple treatments, Belinda devotes herself to helping other people in the community and at the Morgan Hill Bible Church. In their struggle against perchlorate-related illnesses, many of those people spent extra time at the church in counseling and prayer and drank the water there, not knowing that it, too, was highly contaminated by the nearby Olin Corporation.

Olin had actually closed its highway flare manufacturing plant nine years earlier in 1995 after forty years of operation. By 2004, it was paying about $75,000 a month to provide home owners and businesses in the town with 775 gallons of bottled water each day. Most people were still panicky, because they realized that their exposure went far beyond the water they drank.

Sarah Ruby, a tough, talented Stanford graduate who covered the perchlorate problem for *The Pinnacle,* a local newspaper serving the area, reported that a full year earlier, in April 2003, "one family drained their pool to avoid swimming in perchlorate-tainted water. Another pours purified water into their washing machine to rinse her infant's clothes. Another hates bathing her child using the family's polluted well water, knowing that babies end up swallowing bathwater."[38]

Richard Alexander, the attorney for seven local residents who filed a class action lawsuit against Olin, said that the Olin Corporation had known at least fifteen years earlier that their flare-making by-product perchlorate was dangerous to humans, plants, and animals and that they should have known since the early 1960s "that perchlorate was a potential carcinogen that caused developmental disease in babies." But Olin, a $1.5 billion company and the world's largest ammunition producer, claimed that they didn't know. For years, the company had dumped perchlorate into unlined ponds where it percolated directly into the drinking water supply and was consumed by families.[39]

"We operated that facility within all known regulations," said Rick McClure, Olin's spokesman. "We are an environmentally responsible company."[40]

McClure's statements didn't mean much to the Morgan Hill mothers whose babies were born with serious birth defects.

After we talked for a bit, Belinda took me to her church, where I met some of those mothers and spoke with them.

"I started working at the Olin plant in 1974," Virginia Morales said softly as I shared hot dogs, potato chips, and fellowship with her and her daughter, Desiree Hoguin, before evening classes and worship began.

Virginia, a pretty woman with olive skin and a warm smile, spoke about the Olin plant very reluctantly, shyly, and only at my prodding.

"We used to hold the perchlorate in our hands," she said, glancing around the room, then whispering as if, even now, she was still afraid that someone from Olin might overhear her sharing the company's secrets. "We were not required to wear gloves. I was a primer, which means my job was to put the prime inside the flare. I was one of the assembly-line workers. The white powder from perchlorate was always all over me. I feel sure now that they knew much more about the dangers than they ever told us, but they didn't care." Virginia leaned toward me and lowered her voice even more. "I say that because whenever they knew the inspectors were coming, they came through first and opened all the windows. The rest of the time, they kept the windows closed. Usually the perchlorate would be splattered all over the room, practically covering it, but they always had us clean the place up if the inspectors were coming."

"What did it look like?" I asked.

"It was like lumpy dough," she answered, hesitating. "And if you scraped it with something, it would spark. We would put it in big barrels in the back, then leave it there until it was picked up to be dumped somewhere else. We never knew where," she added, dropping her eyes, "because that was kept a secret. I'd come home at the end of each day covered with all the white perchlorate dust, and I'd throw my clothes in the washer with the rest of the family wash. My daughter was exposed to it every day, but, of course, I didn't know then that young people were especially sensitive to it. We lived a couple of miles from the plant, so Desiree, who was just a kid at the time and going through puberty, was not only drinking it in our water and bathing in it and absorbing it through her skin, but she was breathing it, too. That white dust was always all over our house."

"Mom always smelled like burned matches," Desiree said, smiling sadly with her mother's smile. "Like a lot of other girls around here, I developed chronic fatigue syndrome and short-term memory loss. After a while, it got so bad that I'd have to have someone take me to the store because by the time I got there I didn't know why I'd gone. Then when I found out how many people had chronic fatigue syndrome, it began to come together in my mind that something different had happened to us. Of course I still didn't know all the other ways it was affecting us and our unborn children. During my pregnancy, I was there at the plant every day either driving my mother to work, picking her up, bringing her lunch, or all three."

Desiree looked over at her severely disabled son, Aaron, who was quietly playing on the floor beside us, and her eyes filled with tears. "Oh, the guilt, the guilt," she said softly. Then her lips began to tremble as she added, "He was born with velocardiofacial syndrome. That means his face and his whole body are asymmetrical. He was also born with a cleft palate and a hole in his heart. Aaron had Tetrology of Fallot surgery. They had to reconstruct his heart. He had two open-heart surgeries with a valve replacement, and he'll need a third. His growth is so stunted, he's not on the chart. I knew before he was born that something was wrong." Desiree added, her voice rising with emotion, "Six weeks before, I would put my hand on my belly and say, 'This doesn't sound like my child's heart should sound.' In the first moments after he was born, the doctor still didn't know that anything was wrong, but I did. Call it a mother's intuition or whatever, I just kept saying, 'Something is wrong. Something's very wrong.' 'Nothing's wrong,' they said. Then they took him from me and looked at him, and the next thing I heard was that they were waiting for an emergency transport to take him to a hospital in San Jose. I kept asking what was going on, but they wouldn't tell me anything, and they wouldn't let me see him. It took another two days before they told me they were sending him from San Jose to San Francisco, because it was such serious open-heart surgery. Frank Hanley, one of the top pediatric cardiothoracic surgeons in the world, operated on him." Desiree wiped her eyes. "Having a child like this, your whole heart, your whole being, is destroyed.

"I went to the hospital in San Francisco, and I told them that I was staying. I was not leaving without my son. I wanted this child. I wanted him so badly. I was the mother. I felt I had to be there with him. No nurse, no doctor, could take care of my child like I could." Her eyes clouded over for a moment, then took on a glow. "He's definitely been a blessing to the entire family, but it's bittersweet. He's so gentle, so honest. My son has a great, open, loving heart, but it's a heart with a hole in it. I'm anxious about his future.

"Before I became pregnant, I was in graduate school. His father and I thought we would have a son who would graduate from a university. Now, that seems unlikely. Even his life span is uncertain. He may pass away before me. Every time you open up the heart with these surgeries, it does damage. But we're hoping that Aaron will someday become stronger. He's always had great difficulty speaking and swallowing, but until recently he never had any idea that he was different. He's an extrovert, and he's extremely social. That and the sweetness are a blessing."

When it was time for the Bible classes to begin, I said good-bye to Virginia, Desiree, and Aaron, who hugged me even though we were strangers. Then Belinda led me from the dining area into the central hall just in time to catch Kathleen Froess, a young, very pregnant woman with shoulder-length blond hair and bright blue eyes. Belinda had explained to me ahead of time that Kathleen had a badly disabled legally blind hydrocephalic child. And since Kathleen was already part of a perchlorate class action lawsuit, she probably would not be able to speak to me.

I stood back, giving them a chance to discuss it, and was moved to see Kathleen nod yes, then come forward and hold out her hand to me.

"I'm Kathleen Froess," she said in the direct and warm way in which, I was to see, she expressed all things.

I studied her face, amazed that she'd had the courage to become pregnant again. "Can we sit down?" I asked.

She shook her head. "I'd like to, but I don't have much time," she said apologetically. "My Bible study class starts in five minutes."

An hour and a half later we were still standing, riveted to the same spot, still talking.

"At twenty-eight weeks, I found out," she explained calmly. "First, I noticed that the radiologist was spending a long time scanning the baby's brain area. I could tell right away that he looked concerned, but when he said, 'Come back tomorrow and bring your husband,' I knew it was going to be bad.

"The next day, they told us what to expect and offered me an abortion, but I said to myself, 'No, this is your child, and it's in God's hands now.'" Kathleen paused, then breathed deeply. "I'm an occupational therapist and brain injury is my specialty. It's what I had already been prepared to do. I felt that whatever it was, I'd be able to guide my child through it. So far, with help, that's been true. She has been taken care of by some of the best neurosurgeons in the world. Yes, it's been hard, but you know what? I would do it all over again just to have her." Kathleen paused, her eyes filled with tears for a moment, and then the words tumbled out. "She's so funny, so awesome, so wonderful. I've benefited so deeply and gained so much just from knowing her. I suppose I could be mad at God. I could say, 'Why me? Why her? Why this little girl?' " Kathleen shook her head in wonder. "But this child has changed me in ways I never could have imagined.

"I didn't know anything about perchlorate, and I drank our well water every single day when I was pregnant. The night when the attorney came to our house to talk to us and the whole thing sunk in, I thought, 'Oh my God, oh my God. So this is what happened to my baby.' So I joined the lawsuit in the hope that there would be enough money to meet her medical needs."

Kathleen paused again and wiped away the tears that had now run down her cheeks. I just stood there, spellbound by her words and her spirit.

"I think the thing that bothers me the most about perchlorate," she finally said, "is the fact that it damages the fragile citizens most: the elderly, the ill, the pregnant women, the unborn children, and the young children, the ones that the rest of us are charged by God to take care of. They are the very ones whose lives are being poisoned for profit.

"My baby is almost blind. She has fluid around her brain. She's already had a stroke. Her left leg is two inches shorter than her right leg and, because her physical disabilities are so profound, it's hard to tell if her intelligence is normal. But she tries." A sob caught in Kathleen's throat. "Oh God, she tries so hard.

"When she was born, her eyes were bulging down because her head was so swollen. She was only five pounds, four ounces, but her head circumference was forty centimeters, which was really big. We spent almost nine months in the hospital, with seventy-five days in the intensive care unit. We finally took her home at thirty-five weeks. I was completely obsessed with her condition, totally consumed by fear and concern and love. When it's your child, you'll do anything."

Again, I nodded wordlessly. There was nothing I could say.

"She provides a depth to our lives that we never would have had without her. Most parents love their children and would die for their children, but the pain of what happened to her and what she went through touched us deeper, so the love was even more profound. That's the best way I can explain it. At first, nothing else could enter our hearts or minds except the pain and suffering that we were all going through, and then when the love floated up it was a much richer and deeper love than I had ever known.

"So, even with all the problems, I know that we're still lucky. In other countries, I'd have had to let Kristen die. I'm so grateful because here, I can call the hospital and say, 'My child needs you,' and then go there. I am so lucky. I wonder how come I get to keep her and there are other women in other countries and even in this country who don't. A lot of other babies are suffering from birth defects and chemical contamination, and some of them just die."

"So you're living one day at a time with this child and doing whatever you can?" I asked.

Kathleen nodded, then shrugged. "What else can I do?"

"Weren't you afraid to be pregnant again?"

"Yes, I was. But now I have my thyroid function tested all the time. I've been studied and scanned and, so far, this new baby appears normal. And, of course, this time I drink only bottled water."

My mind jumped to an article I had seen in the local newspaper that very morning. "Have you heard about the contaminated milk?" I asked thoughtlessly. "Have you seen today's newspaper?"

"The milk?" Kathleen repeated, looking puzzled.

"Yes. Just today a new report was released. They are finding perchlorate in supermarket milk. The cows are getting it from the alfalfa they eat, because it's irrigated with water from contaminated wells."

Kathleen's eyes widened. She stepped backward and her hand flew across her mouth. "Oh no, no. Oh my God, no," she repeated over and over. "Not the milk. Not the milk. I've been drinking the milk every day."

# 3

# TOXIC FOOD

## Thyroid Disease, Cancer, and Pesticide Victims

> We are men and women from the pews and pulpits of mainstream America . . . loving our Creator and thus caring for God's creation is at the heart of our religious faith . . . We seek to raise our children in a culture of stewardship and bequeath them the full blessings of God's bounty.
>
> —"GOD'S MANDATE: CARE FOR CREATION," NATIONAL RELIGIOUS PARTNERSHIP FOR THE ENVIRONMENT, AN ALLIANCE OF CATHOLIC, PROTESTANT, EVANGELICAL, AND JEWISH ORGANIZATIONS

### PERCHLORATE IN FOOD AND MILK

AS IT TURNED OUT, KATHLEEN'S SECOND BABY WAS BORN NORMAL, but we were still months away from knowing that. When I returned to my hotel room after leaving her, I was angry at myself for having blurted out the findings on perchlorate and milk so carelessly. I reread the June 25, 2004, front-page article in *The Pinnacle*. It confirmed that the Environmental Working Group had just discovered perchlorate in thirty-one of thirty-two milk samples taken from the shelves of grocery stores in the Los Angeles and Orange County areas. Even more significant, I learned that the California Department of Food and Agriculture had completed a study of its own a few weeks earlier, which had not been released to the public. Their study found perchlorate in all thirty-two samples of the milk they had gathered around the state, with contamination levels more than ten times higher than the EPA thought advisable.

I also learned that almost a full year earlier, in September 2003, Texas researchers had done a statewide study and found perchlorate in all seven supermarket samples of milk that they tested.[1]

It was July 2004, and few people had heard that perchlorate collects in foods. Nor did we know that when contaminated water runs through plants, fruits, and vegetables the perchlorate remains there in startlingly heavy concentrations.

The USDA and the EPA had secretly been studying the issue for years. Their findings were eagerly awaited by many people in the field, but in an extraordinary move in May of 2003 the Bush administration placed a gag order on EPA officials, literally banning them from making public what they had learned about perchlorate's impact on the food chain and on human health.

According to Peter Waldman of *The Wall Street Journal,* the White House Office of Management and Budget had turned the perchlorate debate over to the National Academy of Sciences for review and, "pending that study, which could take an additional six to eighteen months, the EPA ordered its scientists and regulators not to speak about perchlorate, said Suzanne Ackerman, an EPA spokeswoman. The gag order prevented EPA scientists from commenting or elaborating on the two lettuce studies."[2] The *Wall Street Journal* article also noted that "the Defense Department refused to fund the roughly $215,000 needed to collect vegetables for sampling," quoting Cornell Long, who heads perchlorate research on food sources for the Air Force. " 'In a perfect world we would have that farm gate data now' on vegetable content, Mr. Long said. 'Everybody thought it was a good idea.' "

According to the same source, "Mr. Long attributed the Pentagon's decision not to fund the study to bureaucratic issues involving budget cycles. Some environmentalists, however, say the

Defense Department simply didn't want to know if perchlorate was in the U.S. food supply because of liability concerns."

The EPA's own study, which was reviewed by *The Wall Street Journal* in the same article, had already shown "that lettuce grown in a greenhouse with perchlorate-contaminated water absorbs and concentrates the chemical at varying rates depending on leaf location. The outer leaves of the lettuce, which the study's authors wrote are usually not eaten, concentrated perchlorate by a factor of seventeen to twenty-eight, meaning the outer leaves contained seventeen to twenty-eight times more perchlorate in them than did the water used to irrigate the plants. The concentration factor for the 'emerging head'—the part people usually eat—was three to nine, the study found.

"Hence, if those results were found to be applicable to winter lettuce grown with Colorado River water, which contains between 3 and 10 parts per billion of perchlorate, the perchlorate concentration in the edible leaves could range as high as 90 ppb." The Environmental Working Group said that level is four times the EPA's recommended daily dose for perchlorate.

Bill Walker and other environmentalists were now deeply concerned, because, as Sarah Ruby pointed out in *The Pinnacle*:

> There still wasn't enough information about perchlorate exposure through vegetable, fruit, milk, and other sources to determine how much people could tolerate in the water supply without becoming ill from the cumulative doses. [The environmentalists] were especially worried about babies, pregnant women, senior citizens, and people with thyroid disease.
>
> Equally disturbing was the fact that a lot of what was already known was still being withheld at the federal level. The [California] Department of Food and Agriculture had

only released their study on perchlorate in milk after the Environmental Working Group heard rumors that such a study existed and filed a public records request. Even then, the Department of Food and Agriculture claimed its test was being taken out of context and said that their perchlorate study was not a public health study. They insisted that it was only meant to evaluate a new perchlorate detection method in food and milk, not to make a statement about the safety of the American food or milk supply.

Bill Walker was incredulous: "Except for ours, this was the largest perchlorate sampling of milk anyone has done anywhere in the world. . . . To say that just because you're not doing a public health study you don't have public health implications is just ridiculous."

Department of Food and Agriculture spokesman Steve Lyle countered by claiming that the nutritional benefit of milk still far outweighed the risk of perchlorate exposure. "California's milk supply is safe, and [people] should continue to consume it. . . . The public health goal for water should not be extrapolated to apply to milk. . . . Milk is different from water."[3]

Indeed, milk is different from water, but, in terms of toxic exposure, no one knew what that meant. By July of 2004, the Department of Food and Agriculture was "estimating that 40 percent of a person's exposure comes from eating and 60 percent from drinking water."[4]

Four months later, on November 30, 2004, WTKR, a local station in Virginia, reported that perchlorate had just been found in Virginia's milk supply. They added that government researchers had also collected five hundred samples of lettuce, milk, and even bottled water from a dozen other states and found perchlorate in virtually all lettuce and milk.

By the next day, the Associated Press and MSNBC.com had picked up the story and elaborated on it. The report they issued made it clear that perchlorate had even been found in organic milk in Maryland and green leaf lettuce in Arizona and, perhaps most striking of all, in bottled spring water in Texas and California. The FDA now acknowledged that the green leaf lettuce grown in El Centro, California, had 71.6 parts per billion of the compound and red leaf lettuce had 52 parts per billion of perchlorate, while whole organic milk in Maryland had 11.3 parts per billion of perchlorate.

When asked whether that level of chemical in organic milk was worrisome, Kevin Mayer, the EPA's regional perchlorate coordinator for Arizona, California, Hawaii, and Nevada, paused, then finally admitted, "The answer is, we don't know yet."

By November 2004, the FDA had admitted finding perchlorate in 93 percent of tested lettuce and milk samples across the nation.

Then EPA official Kevin Mayer called for calm, saying in an interview, "Alarm is not warranted. . . . I think that it is important that the EPA and FDA and other agencies come to some resolution about the toxicity of this chemical. That has been, frankly, a struggle for the last few years."[5]

Larry Ladd is not optimistic that the FDA, the EPA, the Pentagon, industry, or local governments will ever come to any resolution about the dangers of perchlorate or ever tell the public what they already know about safe levels.

"In Rancho Cordova," he explained, "we waited seven years for the 'first step' in the evaluation of our perchlorate exposure. It is an infant thyroid hormone study. It was literally weeks from completion in August 2001. Three years later, in 2004, California's Office of Environmental Investigations claimed that they were still 'working' on it.

"The truth is," Larry added angrily, "that they don't want us to know what they found out, because then they would have to try to fix it. Industry usually knows years before the public about environmental contamination problems, but regulators rarely pin direct blame for poisoning people through water, food, or air on any industry. At best, they regulate a chemical to limit future poisonings and continue to pretend they have done nothing wrong."[6]

"If you really want to know why they are hiding the perchlorate results, I'll tell you. It's because they are terrified of the chaos and national panic that would result. You are about to meet living proof of where we are all headed, a family who found out firsthand, the hard way, what massive doses of perchlorate in the water and food chain can do."

Larry was speaking to me as we drove back from Morgan Hill and parked across the street from the modest ranch-style home of Greg and Doris Voetsch, located in Rancho Cordova just fifteen miles from downtown Sacramento. I was aware that the Voetsches had moved to this house in 1970, but I did not yet know that Mr. Voetsch had had two cancerous tumors removed from his thyroid gland in 1983. Nor did I know that both of his daughters, who were in their twenties, had required surgery to treat thyroid-related problems or that in 2000 his wife, Doris, had two breast cancers and a benign brain tumor removed. To make things worse, in 2004, coincidentally on the day that I visited, their forty-four-year-old son was diagnosed with schwannoma, a rare thyroid tumor the size of a grapefruit that had grown in his chest, shoulder, and back. Multiple small tumors surrounded it. The family's exposure to perchlorate combined with NDMA, another suspected carcinogen that has also been found in Rancho Cordova's drinking water, was particularly intense because Greg Voetsch was a farmer. In addition to drinking the contaminated water, for years the Voetsches had eaten homegrown tomatoes, lettuce, brussels sprouts, carrots, radishes, cucumbers, string beans, apples, pears, apricots, and figs, all watered from their perchlorate-contaminated well. They also ate salmon and rainbow trout caught in the perchlorate-poisoned American River flowing just beyond their backyard at a time when the chemical was measured at levels reaching 300 parts per billion.

"I didn't know what was going on," Mr. Voetsch told me as he sat in his living room, drinking a glass of bottled water and hoping it was safe. "I had no idea. I just sat there in my garden, watering it with contaminated water and eating the food right off the vine. You see, I loved growing my own food. I thought it was healthier. That was my downfall. By 1992 or 1993, I was so sick I couldn't even get out into the backyard to pick the food anymore.

"By that time, I knew something had to be wrong because the whole family was sick. I just didn't know what. I had four trees in the yard that suddenly died. The cherry tree died first, then the apple tree, then the apricot tree, and finally the fig tree. Two of them had blossomed out in the spring and were all set for fruit when suddenly the blossoms left and the trees died. That's not normal. These were new trees that I'd just planted.

"When I heard on the news about a thyroid-disrupting contaminant that was shutting down all our wells, poisoning our trees, our food, and our people, I thought, 'Holy Mother of God, so that's why my trees died, and that's why my entire family has thyroid problems. I've been poisoning them.' I began making calls. I called all the county offices I could think of, and I called the University of California at Davis and a bunch of other places, but not one of them ever called back. No one gives a damn until someone dies. The only person who returned my calls was Larry Ladd."

Larry nodded, then said, "It's time our children quit drinking and eating rocket fuel. Greg joined a class action lawsuit, but it was stuck in the courts for years. It wasn't until after *The Wall Street Journal* came to Rancho Cordova and interviewed Greg

Voetsch that things began to change. Maybe it was just a coincidence, but the weekend following the *Wall Street Journal*'s article on perchlorate, Christine Whitman resigned as head of the EPA. Then there was a TV interview with Greg right after that on KPIX TV, which is part of CBS. Aerojet actually tried to confiscate the tape of that interview right in the parking lot, but the cameraman managed to slip away. About a week after that, Greg was attacked and nearly killed."

"Was it related?" I asked.

Larry and Greg exchanged glances; then Greg shrugged. "A lot of people think so, but we can't prove the guy was definitely targeting me. I was driving along," he explained. "Suddenly, out of nowhere, this guy came up in front of me and gave me the finger. He nearly pushed me off the road. At the next light, he cut me off again. Then, when the light changed, his lane got bogged down and mine was still moving, so, naturally, I passed him. He caught up with me and then cut me off for a third time. So then, like an idiot, I got out of my car at the next light and walked over to him and I said, 'What the hell is wrong with you?' That's the last thing I remember. Witnesses say I got four or five rapid karate punches. Then, when I fell, the guy kept kicking me in the face. Because of the fractures, I lost vision in one eye. I'm supposed to have surgery to try to restore it.

"My doctor at the VA hospital had warned me about the possibility of retribution before I went public, but, you know what, at that time I didn't believe such a thing was possible in the United States of America. Still, I'm not sorry I spoke up. My family and I have lived in Rancho Cordova for over thirty years, right on top of one of the heaviest concentrations of perchlorate that's ever been discovered. Aerojet knew about it for years. They moved all their employees to the other side of the river in an effort to keep them safe, and that makes me really mad, not so much for me, but for my wife and my son and my daughters. Every one of us is sick with perchlorate-related illnesses. I'm a farmer and a landscape contractor and so is my son. We don't have a lot of money. Maybe we're just little guys whose lives don't count, but you know what, we love our country and we don't want to see it contaminated. Now that we know what's happening, we feel we have an obligation to try to protect other people so that what happened to us doesn't happen to them."

Greg took another sip of his bottled water. Larry smiled warmly, then added, "About an hour after Greg's appearance on the evening news, more than seven hundred people gathered at a town meeting in Morgan Hill to ask questions about their own plume of perchlorate. They were panicky because it was creeping across San Martin and into Gilroy, poisoning at least eight hundred domestic wells and municipal water systems. After watching Greg on TV and hearing his story, people asked a lot of frantic questions about their health, their livestock, and even their real estate values because, suddenly, everybody knew that nothing would ever be the same again."

Shortly before this book went to press, Greg Voetsch called to give me an update. After four surgeries, his eyesight had still not been restored to normal, but Greg was not focused on that. His wife's cancer had returned in both breasts. After Doris underwent two mastectomies, doctors did a CAT scan and discovered lung cancer. They also performed a colonoscopy and found advanced colon cancer.

"We are in unchartered waters," the doctors said. "We've never dealt with three primary cancers at once before."

Aerojet made an out-of-court settlement with the Voetsches.

"It was a lot of money," Greg said. "But it didn't bring back our health. They admitted no guilt and the settlement terms don't allow me to reveal any of the details."

"Did it change your lives?" I asked.

"Well, I put new siding on the house," Greg answered. There

was a pause; then he added, "I would put every penny back if we didn't have to go through this. My wife is a pillar of strength. We are strong in the church and that has helped us get through this. It's been hard, very hard, but, because we have the support of the Lord, it has not been devastating. I plan to spend my remaining time making sure people know just how deadly perchlorate can be."

According to *The Pinnacle,* Brian Leahy, president of the California Certified Organic Farmers in Santa Cruz, said, "We all feel violated. Farmers doing the right thing are going to be hurt by perchlorate. Twenty million people in Southern California are drinking this accident, and there is no covering it up. Chemicals you dump down the drain today will end up in your grandchildren's bodies tomorrow. You can't hide from these things once you put them in the environment."[7]

The situation is even more urgent than Brian Leahy indicated. That's because eating organically grown food makes little difference if that food is irrigated with perchlorate-contaminated water. Perchlorate is then passed through the water supply into the organic foods and ultimately into the breast milk. Without their knowledge, mothers all over America and beyond are already passing perchlorate from both organic and conventional food to their babies through their contaminated breast milk.

A study published by Texas Tech University on February 22, 2005, found dangerous levels of perchlorate in every one of the thirty-six samples of breast milk in nursing mothers in all eighteen states tested. Renee Sharp, a senior analyst at the Environmental Working Group, told Marla Cone of the *Los Angeles Times,* "The findings of today's breast milk study raise serious concerns for breast-feeding infants.

"This is not just another study," Sharp added. "It definitely ends the questions about whether women are passing along perchlorate to their kids through breast milk, and the sky-high levels the scientists found put more than half America's kids over the levels that even the National Academy of Sciences is recommending as safe."[8]

Texas Tech University, which is recognized as the nation's top perchlorate-testing laboratory, was the first ever to look for perchlorate in human breast milk. The average levels they found were 10.5 parts per billion, five times higher than the average level found in forty-seven samples of cow's milk collected by the same scientists from supermarkets in eleven states. The levels of perchlorate in breast milk were more than twice as high as the levels found in the combined total of 222 samples of cow's milk from twenty-one states that had been tested for perchlorate by the FDA, the state of California, Texas Tech, and the Environmental Working Group.

Scientists had long feared that perchlorate could end up in and perhaps even be concentrated in breast milk. But until February 22, 2005, they didn't have any proof. The Texas Tech study not only showed that the contamination existed, but it also proved that the contamination was pervasive and that the levels were higher than anyone ever expected. "Five women had breast milk with perchlorate concentrations over 20 parts per billion, and one woman's breast milk had a concentration of 92 parts per billion. The highest levels were found in women from New Jersey, New Mexico, Missouri, Nebraska, and California, in that order."[9]

The finding surprised toxicologists, because unlike many other industrial chemicals, perchlorate in breast milk does not build up in tissues over time. Instead, the amount passed on to the infant is determined by what the mother has just eaten.[10]

As the Environmental Working Group asserted, "After so many years of deliberate delay in setting safety standards and cleaning up perchlorate-contaminated sites, the high level of perchlorate found in mother's breast milk should be more than a wake-up call to state regulators, the EPA, the Department of Defense, and the Bush administration."[11] It should be a mandate for change.

Perchlorate from breast milk is just one of many kinds of contamination that are passed from mothers to children. According to a benchmark study released in July 2005 by the Environmental Working Group, mothers actually pass on hundreds of other dangerous chemicals through their umbilical cords. The Working Group's study found 287 industrial chemicals and pollutants in the umbilical cord blood taken from a sample of ten babies born in U.S. hospitals. The blood harbored pesticides, flame-retardants, chemicals from nonstick cooking pans and plastic wrap, as well as long-banned PCBs and wastes from burning coal, gasoline, and garbage. On average, each baby had been exposed to 200 chemicals. These exposures through the umbilical cord are nearly impossible for mothers to control or avoid. This study was the first to measure a baby's "body burden," the term used to describe the amount of pollutants buried in the bloodstream, organ tissues, and fat cells of newborns.

"Of the 287 chemicals detected, 180 were suspected carcinogens, 217 were toxic to the brain and nervous system, and 208 have been linked to abnormal development and birth defects in lab animals."[12]

"The numbers are startling," said Dr. Alan Greene, a Stanford University pediatrician who wrote a commentary on the study.

In the month leading up to birth, the mother's umbilical cord pumps at least three hundred quarts of blood back and forth each day from the placenta to the fetus, bringing the baby oxygen and nutrients. Scientists once believed that the placenta shielded the cord blood and the baby from most chemicals and pollutants. Unfortunately, they were wrong.

Dr. Greene described the placenta as a "free-flowing, living lake" from which the blood vessels in the umbilical cord draw. He said, "Today, this most primal of lakes has become polluted with industrial contaminants, and developing babies are nourished exclusively from this polluted pool."[13]

"Our wombs are no place for poisons. Our babies have the right to be born toxic-free," said Laurie Valeriano, policy director of the Washington Toxics Coalition and mother of three. "It's time for a complete overhaul of the current system," she added. "Government should phase out very harmful chemicals and industry must substitute safer substances when they are available. There is no reason consumer products should be filled with chemicals that poison babies when there are safer alternatives."

Physicians for Social Responsibility added, "It is clear that the developing fetus, infants, and young children are particularly sensitive to the harmful effects of pesticides." Another study concluded, "Effects of exposure during development are permanent and irreversible."[14]

## METHYL BROMIDE

One of the most dangerous pesticides is methyl bromide. In December 2005, Rita Beamish of the Associated Press wrote, "As American shoppers rifle through shelves brimming with succulent tomatoes and plump strawberries . . . there is no hint of a dark side. Other nations watch as the United States keeps permitting wide use of methyl bromide for tomatoes, strawberries, peppers, Christmas trees, and other crops, even though it signed an international treaty banning all but the most critical uses by 2005."[15]

As the year turned to 2006, the Bush administration, at the urging of agriculture and manufacturing interests, was making plans to ensure that the extremely lucrative methyl bromide remained available at least through 2008 by seeking "critical use" exemptions. The administration would not commit to an end date.

Beamish explained: "Odorless and colorless, methyl bromide is a gas that is usually injected by tractor into the soil before planting, then covered with plastic sheeting to slow the chemical's release into the air." Methyl bromide leaves an abundant crop by destroying everything else in its path: parasites, disease, weeds, and sometimes people. Those who inhale large amounts of the chemical can experience convulsions, neuromuscular and cognitive problems, coma, and death.

According to Kristin Collins writing in *The News & Observer,* "One worker told inspectors that he sometimes worked in the fields while methyl bromide was being applied. By law, workers cannot reenter a field until forty-eight hours after the application. . . . The worker, Oscar Hernandez, said in a telephone interview that during the sixteen months he had worked for Ag-Mart tractors often spayed fields without warning while he was collecting debris and working on irrigation systems in them. He also worked as a pesticide handler and said he received no training for the job. Hernandez, who is thirty-six, said through a translator that he sometimes felt dizzy and agitated, almost drugged, while pesticides were being sprayed. He said he still suffers from headaches, nervousness, and memory loss, all documented effects of pesticide exposure."[16]

The Environmental Working Group calls methyl bromide "an extraordinarily toxic poison" and classifies it as "a category one acute toxin," the most potent class of toxic chemical that exists. They report eighteen known deaths from acute poisoning and more than 454 people who became sick and had to leave schools and homes as a result of exposure.[17] The actual numbers are thought to be far higher, because people often do not seek medical care, and even when they do, the illness/pesticide connection usually remains undocumented.

*The Palm Beach Post* reported:

> Carlos Candelario, known as Carlitos, was born December 17 without arms or legs. On February 4, Jesus Navarrete, whose parents live about one hundred feet away from

Carlitos's family, was born with Pierre Robin syndrome. His jaw is underdeveloped, and that causes his tongue to fall into his throat, and he risks choking. Two days later, on February 6, Maria Meza gave birth to a child missing its nose and an ear and with no visible sexual organs. At first the child was given the name Jorge, but hours later was renamed Violeta after a more detailed examination determined that the baby was a girl. She died three days later of massive birth defects. . . .

In 2004 when they became pregnant, all three mothers lived within 200 feet of one another at the same migrant labor camp called Tower Cabins. All of them are Mexicans who worked for the same produce company picking tomatoes. . . . When the harvest was completed in Immokalee, they moved on to fields in North Florida and North Carolina, but they say they continued to work for the same employer and were again exposed to agricultural chemicals. . . .

"People have mentioned to me that maybe this has to do with chemicals," said Francisca Herrera, nineteen, Carlitos's mother. "But I really don't know anything about that. I would like to know. . . . When you work on the plants, you smell the chemicals."

"It has happened to me various times that when you are working and the chemical has dried and turned to dust, you breathe it," said Meza. "They say it's dangerous." . . .

Herrera said her son, Carlitos, is acting normal despite having no arms and legs. "He eats well and he sleeps well and I think he is intelligent," she said proudly.[18]

"The couple were unaware of Carlitos's deformities until his birth. . . . It wasn't until after the child's birth that Herrera learned of the deformities. 'They asked me if I worked in the fields,' she said. 'And I said yes.' "[19]

In another *Palm Beach Post* article Don Long, president of Ag-Mart, issued a public statement that said in part, "We are deeply saddened by what Carlitos and his family have endured over the past year; it's a heartbreaking experience for any family. From the beginning, we cooperated fully with authorities, conducting two independent investigations into this case. . . . We continue to cooperate with any agency still investigating this issue."[20]

While the investigations continue, many of us will keep eating and breathing methyl bromide and we will go on hanging our holiday ornaments on trees that have been treated with the chemical. The impact of long-term low-level exposure to it is still unknown. But logic suggests that methyl bromide, like all the others, will end up entering our bodies and being passed on to our unborn children.

The Associated Press reported that the family of Floyd Gottwald, vice chairman of the Albemarle Corporation of Richmond, Virginia, a methyl bromide producer, have contributed more than $420,000 to President Bush's campaign and to Republican Party organizations since 1999.[21] The EPA has refused to say what the U.S. inventory of methyl bromide is, and a company spokesman said Ag-Mart intends to continue using methyl bromide.

# 4

# THE CYCLE OF AIRBORNE POISON

## Mercury Contamination and Asthma

> The whole human race suffers as a result of environmental blight, and generations yet unborn will bear the cost for our failure to act today.
>
> —THE U.S. CONFERENCE OF CATHOLIC BISHOPS

### MERCURY IN FOOD

NEWS OF METHYL BROMIDE AND THE MASSIVE CHEMICAL CONtamination of all newborns by over two hundred industrial pollutants, carcinogens, and pesticides came on the heels of an earlier shock. In February 2004 the EPA had issued an unexpected warning about mercury levels in newborn babies that had grave implications.

The announcement acknowledged that infants' umbilical cord blood contained much higher mercury concentrations than their mothers' blood. The EPA admitted that twice as many babies as they had previously thought were being exposed to extremely dangerous levels of mercury in the womb. About 630,000, or one in every six babies born each year in the United States alone, were at risk.

This toxic mercury originates in coal-fired plants. First, it pollutes the surrounding air, then, in an inexorable cycle, it drifts with the wind and rains down into the ocean water. Once it has settled in the sea, it begins contaminating the smallest fish. But it quickly moves through the food chain from smaller to larger fish, intensifying in contamination concentration all along the way, with the largest fish, like tuna, holding the highest mercury levels. Ultimately, the newborns' exposure to this form of mercury is from the fish and shellfish their mothers eat.

As Sandra Steingraber, the noted scientist and author of *Living Downstream,* points out, ancient anaerobic bacteria found in the marine sediments transform this heavy metal into methyl mercury. Since tuna is a top-of-the-food-chain predator, the methyl mercury concentrates in the flesh of the tuna's muscle tissue.

According to Steingraber, "There is no 'organic' option for buying tuna . . . no special way of cleaning or cooking tuna that would lower its body burden. Nor is there a way of . . . preventing those molecules of mercury from interfering with brain cell functioning. In that sense, the problem of tuna fish is more akin to the problem of air and water pollution. It is not a problem we can shop our way out of."[1]

The EPA's toxics and pesticides office announced that because the "science" of exactly how much mercury contamination travels from coal-fired plants to fish to baby's blood is still evolving, their estimates of "safe" amounts of mercury from contaminated fish consumption might have to be changed again.

Sandra Steingraber added that "its presence in umbilical cord blood is especially troubling because methyl mercury has been shown to paralyze migrating fetal brain cells and halt their cell division. As a result, the architecture of the brain is subtly altered in ways that can lead to learning disabilities, delayed mental development, and shortened attention spans in later childhood."[2]

As a result of the EPA's findings, the FDA revised its recommendations to American women of childbearing age about the quantity and type of seafood they should consume to avoid damaging their babies' mental and neurological development. In March 2004, for the first time, the FDA added tuna to the list of fish to be avoided. The FDA had previously urged women who planned to become pregnant to choose fish that had lower mercury levels and to avoid swordfish, shark, tilefish, and king mackerel since they were especially high in mercury.

According to the *San Francisco Chronicle,* the tuna advisory had been "long awaited." It had finally come from the FDA and the EPA after months of intense controversy and compromise. "Although the recommendations have no binding force, they carry enormous weight as a guide to states, physicians, nutritionists, and the public on how to best control mercury in the food supply. . . . Environmental health advocacy organizations said the new guidelines don't go nearly far enough in warning consumers of the mercury danger."[3]

According to the same source, at that point the FDA "advisory sets a weekly allowance on albacore tuna at six ounces, or roughly one standard-size can, for all those considered sensitive to mercury, including children, pregnant and nursing women, or any woman who may become pregnant. Children are cautioned to eat smaller portions than adults, but no specific amounts were included in the guidelines."

One leading FDA advisory panel expert resigned in protest "after learning that the FDA was going to 'disregard' science" and not warn consumers about the real health risks of eating tuna. "University of Arizona toxicologist Vas Aposhian said the advisory should have put more stringent limits on all canned tuna. . . . The new recommendations are dangerous to 99 percent of pregnant women and their unborn children. . . . 'It seems that one should be more concerned about the health of the future children of this country,' he said, 'than the albacore tuna industry.' "[4]

The U.S. Tuna Foundation, which represents StarKist, Chicken of the Sea, and other canned tuna producers, issued a statement of their own applauding the FDA for emphasizing that "consumption advice was not necessary for the general population."

Other critics charged that albacore mercury levels were so high that the six-ounce recommended weekly allowance would expose fetuses, infants, toddlers, and growing young people to levels many times higher than the EPA's safety guideline.

To get back to the *Chronicle* article: " 'If American women follow the FDA's advice and eat a can of albacore tuna a week, a bad situation will be made far, far worse,' said Richard Wiles, senior vice president of the Environmental Working Group in Washington. He said the advisory 'flies in the face of all scientific understanding of the hazards of mercury to children.'

"A San Francisco pediatrician, Dr. Michelle Pepitone, expressed disappointment that the advisory wasn't more protective. 'The government is shirking its duty to protect our children. I'm shocked and saddened,' she said."[5]

According to independent testing conducted for the Mercury Policy Project,[6] a center for science in the public interest based in Vermont that works to reduce mercury exposures at the local, national, and international levels, the tuna industry and the FDA had both known of tuna's high mercury content for years. The Mercury Project's own test results indicated that "one out of every twenty cans of 'white,' or albacore, tuna should be recalled as unsafe for human consumption." Michael Bender, the director of the project since 1998, said that the levels of mercury in white tuna on average were "considerably higher" than the industry and government had ever admitted.

"We chose sixty cans of tuna randomly off grocery shelves," Bender explained. "We had them tested . . . and then had a portion retested by the National Food Laboratory, Inc.—a lab used by the tuna industry—so there is no reason to believe that these results are not reflective of what millions of Americans consume."

Test results from the FDA not only confirmed the Mercury Policy Project findings, they also showed that white canned tuna had three times the mercury levels of the light tuna. According to same article, "Canned tuna is consumed in 90 percent of American households and accounts for around 20 percent of U.S. seafood consumption. Children eat more than twice as much tuna as any other fish, and canned tuna is the most frequently consumed fish among women of child-bearing age. Albacore accounts for about one-third of all canned tuna sold in the U.S."

Peter Waldman of *The Wall Street Journal* pointed out that "the FDA had known for many years that canned tuna contained mercury." Consumer groups had long urged the agency to address the issue. "But it wasn't until March 2004, after regulatory tussles between health advocates and the tuna industry and between clashing scientists for the FDA and EPA, that those agencies issued a mercury advisory that cited tuna." Even then, it was only after the EPA's top mercury risk assessor, Kathryn Mahaffey, warned that "fetuses concentrate more mercury in their blood than do their pregnant mothers."[7]

According to Waldman, "Some EPA scientists insisted that FDA officials were coddling food companies. 'They really consider the fish industry to be their clients, rather than the U.S. public,' charges Deborah Rice, a former EPA toxicologist." The FDA publicly denied that "commercial concerns played role in the agency's decision making."

History suggests otherwise. Late in 2003, FDA and EPA officials proposed their first joint mercury advisory. It was during a hearing of the FDA's Food Advisory Committee. According to the official transcript of the meeting, quoted by Waldman, FDA scientists said they had "put fish in three categories: high in mercury, medium, and low. The level for the low-mercury group was that of canned light tuna." That appeared to be a business decision. FDA official Clark Carrington admitted, "In order to keep the market share at a reasonable level, we felt like we had to keep light tuna in the low mercury group."

Dr. David Acheson, director of food safety and security at the FDA, "told the meeting the fish categories 'were arbitrarily chosen to put light tuna in the low category.' " Dr. Rice responded by saying, "Here's the FDA making what are supposed to be scientific decisions on the basis of market share. What else is there to say?"

On June 13, 2006, the highly regarded Consumers Union issued their own recommendation that pregnant women avoid eating tuna of any kind. "The recommendation came after the *Chicago Tribune* 'reported that about 15 percent of canned light tuna—the variety touted by the FDA as 'low mercury' and thus safe for children and pregnant women to eat in moderation—is culled from species with high concentrations of mercury.

" 'This is important information that women need to hear,' says Consumers Union food policy director Jean Halloran, who also serves on an FDA advisory panel on mercury in seafood. 'We think that high exposures, even for a day or two, could be too much of a risk,' she added."[8]

Food processors had lobbied the Bush administration, asking officials at the White House not to mention tuna. "They said doing so would only drive people, especially the poor, to eat more junk food."[9]

In fact, it isn't just the poor who consume large quantities of tuna and other fish high in mercury. A survey of Bay Area residents who were wealthy and seeking health advantages by consuming fish discovered that they were being adversely affected by high levels of mercury.[10] Nearly all those tested who had been eating fish high on the food chain had excessive levels of mercury in their bodies. The study was the first of its kind. "We found that if people eat fish, the mercury goes up. If they stop eating fish, the mercury goes down. It's that simple," said the author of the study, Dr. Jane M. Hightower.

Hightower tested 123 patients who she knew were consuming a lot of fish. Many were complaining of headaches, memory loss, pains in their joints, and fatigue. "Eighty-nine percent had mercury levels exceeding the 5 parts per billion level recognized as safe by the EPA and the National Academy of Sciences. Sixty-three people had blood mercury levels more than twice the recommended level, and nineteen had blood levels four times the level considered safe. Four people had levels greater than ten times as high as the government recommends."

The study looked at seven children. "All had too much mercury except one who didn't eat fish. One child had a blood mercury level of 13 parts per billion. She was eating two cans of tuna a week, within the guidelines recommended by the FDA. Her mother reported that she was lethargic, lost verbal skills, and forgot how to tie her shoes.

"Hightower instructed her ill and high-mercury patients to give up fish for six months or eat fish that doesn't accumulate mercury, such as salmon, sardines, sole, tilapia, or small shellfish. Mercury levels fell dramatically in the sixty-seven patients she followed in the study."

One patient, Susie Piallat, had suffered from flu symptoms for years. Dr. Hightower found that her mercury level was fifteen times higher than the amount considered safe.

The article quotes Piallat, a former Pan American Airlines sales manager: "I've been very health-conscious all my life. I read everything on the latest medical issues. I'd been eating eight or nine servings a week of tuna, swordfish, halibut, and sea bass, and loading up on mercury. It took almost a year for my [mercury] level to drop. Now I feel so much better."

Dr. Hightower believes that the public needs to be warned about mercury and that advisories should be posted wherever fish is sold.

The parents of a young boy whom Dr. Hightower treated agree with that assessment. Starting in late 2002, their son, ten-year-old

Matthew Davis, ate a tuna fish sandwich almost every day for a year. When Matthew was in fifth grade, his mother, Joan Ellen Davis, requested a meeting with his teachers to find out if they were noticing the same things in school that she was seeing at home.

As Peter Waldman reported in *The Washington Post,* "One by one Matthew Davis's fifth-grade teachers went around the table describing the ten-year-old boy. He wasn't focused in class and often missed assignments, they said." Even the most basic math was suddenly a struggle. His mother added that she had noticed that his fingers looked deformed. He was having trouble playing the guitar, and he couldn't even catch a ball. Tests determined that Matthew had mercury poisoning. "The doctors' prescription was simple: Matthew should stop eating canned tuna."[11] The same article continues:

> Based on the FDA's data for canned albacore, he was consuming a daily dose of mercury at least twelve times what the EPA considered a safe level for a sixty-pound child. . . .
>
> Like many parents, the Davises in San Francisco always thought fish was great. They knew it was high in omega-3 fatty acids, which they understood could help brain development. They were delighted when Matthew started eating what Ms. Davis called "his brain food" for lunch and snacks. . . .
>
> Then Matthew's father happened to read an article in the *San Francisco Chronicle* describing adults with similar problems as a possible result of eating too much swordfish, tuna steaks, and other high-end fish in restaurants. Mr. Davis remembers bolting to the pantry and throwing away eight pouches and twenty cans of albacore tuna. . . .
>
> Nearly two years after Matthew quit eating albacore tuna, his blood mercury level is zero and his condition is dramatically improved.

As Sandra Steingraber pointed out in an article that appeared in *Orion* magazine, "Children do not want to eat a food they like once a month or even once a week. When children discover a new food item to their liking, they want it all the time. They want it for breakfast, lunch and dinner from here to Sunday. . . . How then do you explain to a young child . . . that she'll have to wait until next month before she can have her favorite dish again? Do you tell her that she's already consumed her monthly quota of a known brain poison as determined by the federal government? Or do you make up some other excuse?"[12]

Even limited to albacore, the tuna advisory issued by the FDA caused sales to drop 10 percent, causing a revenue loss beginning in March 2004 of $150 million. By August 2005, the industry was trying to raise $25 million a year to fund an advertising campaign designed to boost tuna sales. If approved by the Office of Management and Budget, which regulates such marketing efforts, the TV advertising campaign, tentatively called "Tuna, a smart catch," which had test runs in Pittsburgh, Pennsylvania, and St. Louis, Missouri, could soon begin airing in American households.[13]

The advertisements would ignore the mercury issue and highlight the various health benefits of tuna fish.

"It's got these great omega 3s," said John Stiker, an executive vice president at Bumble Bee, the nation's number-two tuna producer behind StarKist. The campaign would use the same Department of Agriculture checkoff programs that have financed the "Got Milk?" campaign and the "Beef. It's What's for Dinner" ads.[14]

Nowhere is it clearer than with the mercury contamination of children that money and politics directly affect our health. Most doctors don't test for it and most American women are still unaware of the mercury levels in their own blood.

In the winter of 2005, when Europe and other industrialized regions were grappling with the difficult problem of controlling mercury pollution from all sources, the United States was once again officially downplaying the danger. The Bush administration had already sided with the coal-fired utilities in developing rules for toxic mercury pollution.

On February 3, 2005, the EPA's inspector general, whose job it is to oversee the agency's work on mercury, charged that the EPA had based its mercury pollution limits on an analysis submitted by a group representing seventeen coal-fired utilities in eight western states called Western Energy Supply and Transmission Associates. He also said that the agency's senior management had instructed staff members at the EPA to arrive at a "predetermined conclusion favoring industry." The fifty-four-page report cited anonymous agency staff members and internal e-mail messages, then concluded that the technological and scientific analysis by the EPA had again been "compromised," this time to reduce cleanup costs for the utility industry.[15]

By coincidence, later that same week, in February 2005, the Associated Press reported that speakers at the national meeting of the American Association for the Advancement of Science were expressing their own concerns that scientists in other key federal agencies were either "being ignored or pressured to change other environmental study conclusions that didn't support policy positions. . . . Rosina Bierbaum, dean of the University of Michigan School of Natural Resources and Environment, said the Bush administration has cut scientists out of some policy-making processes, particularly on environmental issues. 'In previous administrations, scientists were always at the table when regulations were being developed. Science never had the last voice, but it had a voice.' . . . Proven, widely accepted research is being ignored or disputed, she said."[16]

Kurt Gottfried, of Cornell University, is quoted in the same article: "A survey of scientists in the U.S. Fish and Wildlife Service found that about 42 percent felt pressured not to publicly report any findings that did not agree with Bush policies. They were even pressured not to express any views in conflict with the Bush policies within the agency. 'This administration has distanced itself from scientific information,' said Gottfried. He said this is part of a larger effort to let politics dominate pure science. Scientists in the EPA have been pressured to change their research to keep it consistent with the Bush political position on environmental issues. Because of such actions, it has become more difficult for federal agencies to attract and retain top scientific talent. This becomes a critical issue because about 35 percent of EPA scientists will retire soon, and the Bush administration can then 'mold the staff' of the agency through the hiring process."

Here is what Julie Cart, writing in the *Los Angeles Times,* had to say:

> More than 200 scientists employed by the U.S. Fish and Wildlife Service say they have been directed to alter official findings. . . . More than half of the biologists and other researchers who responded to the survey said they knew of cases in which commercial interests . . . had applied political pressure to reverse scientific conclusions deemed harmful to their business. . . .
>
> One scientist working in the Pacific region, which includes California, wrote, "I have been through the reversal of two listing decisions due to political pressure. Science was ignored—and worse, manipulated—to build a bogus rationale for reversal of these listing decisions."
>
> More than 20 percent of survey responders reported that they had been "directed to inappropriately exclude or alter technical information." . . .
>
> A biologist in Alaska wrote, "It is one thing for the department to dismiss our recommendations. It is quite another to be forced (under veiled threat of removal) to say something that is counter to our best professional judgment."[17]

None of this was new. During the 2004 presidential campaign, hundreds of scientists, including at least twenty Nobel laureates, had signed a petition organized by the Union of Concerned Scientists, an advocacy group based in Cambridge, Massachusetts, charging the Bush administration with distorting scientific information to advance its political agenda.

In February 2005, the entire European Union was open to a global treaty to control mercury but the United States rejected the idea. The Bush administration suggested "voluntary partnerships," which would leave mercury reduction entirely up to the industries.

One month later, in March 2005, when the EPA finally announced its very limited "cap and trade" rule, American officials claimed that controls to reduce mercury should not be more aggressive because the cost to industry already far exceeded the public health payoff.

An investigation by *The Washington Post* later revealed that a Harvard University study had already reached exactly the opposite conclusion. The Harvard analysis paid for by the EPA, coauthored by an EPA scientist, and reviewed by two other EPA scientists actually estimated that the health benefits of removing mercury were 100 times greater than the EPA had admitted. Once again, these EPA findings had been deleted from the public documents. The Harvard study concluded that controlling mercury could save nearly $5 billion a year through reduced health problems. After deleting this study, EPA officials announced that the health benefits of more stringent mercury controls were worth no more than $50 million a year and that the cost to industry would be $750 million a year.[18]

On paper, the "cap and trade" program that the EPA endorsed would satisfy a regional "cap" on mercury and meet general requirements. The administration said this would provide the utilities with a way to meet an overall industry goal, without spending money on expensive new scrubbers. Power plants that reduced their pollution to below government-imposed caps would have excess pollution "credits" that they could sell to other companies. This would create "hot spots," entire neighborhoods whose inhabitants would continue to be irrevocably harmed. They would be harmed first through the air they breathed as mercury made its way from the power plants toward the nation's rivers, streams, lakes, and oceans, and then through the consumption of tuna and other mercury-poisoned fish.

## AIRBORNE MERCURY AND ASTHMA

In a speech given at the Commonwealth Club, Robert Kennedy, Jr., spoke of the specific damage that airborne mercury caused to the lungs and airways of children. "Consider just one industry: the coal-burning utilities," Kennedy said. "One out of every four black children in [Harlem,] New York now has asthma." Then he added, "I have three sons who have asthma. We don't know why we have this epidemic of pediatric asthma, but we do know that asthma attacks are caused primarily by two components of air pollution: ozone and particulates. . . . We know that the principal source of ozone and particulates in our air is coming from 1,100 coal-burning power plants that are burning coal illegally. They were supposed to install controls over fifteen years ago. The Clinton administration was prosecuting seventy-five of the worst of those plants. But [this] industry gave $48 million to President Bush during the 2000 campaign, and they've contributed $58 million since. One of the first things that President Bush did when he came to office was to order the Justice Department to drop all seventy-five of those suits." Kennedy went on to say that "the Justice Department lawyers were shocked that a presidential candidate had accepted money from a criminal and then let that criminal go free. We are living today in a science fiction nightmare, a world where, because somebody gave money to a politician, our children are brought into a world where the air is too poisonous for them to breathe."[19]

A study conducted in 2003 by Harlem Hospital Center and Harlem's Children Zone confirms Kennedy's assertion that "one of every four children in central Harlem has asthma." This is one of the highest rates ever documented in America. "The Harlem Hospital findings suggest that if blanket testing were more widespread, 'the rates might be much higher than suspected in any number of inner-city neighborhoods around the country.' "[20]

According to the *Village Voice,* an EPA rule signed on August 27, 2003, "allows the nation's most polluting power plants to upgrade equipment without implementing new emission control measures." Because many of those power plants are in the airborne contamination path of low-income areas, "the rule will likely have a disproportionate impact in such neighborhoods," said Judith Enck, policy adviser to then State Attorney General Eliot Spitzer. She noted that "air pollution has been cited as one cause of skyrocketing asthma rates in Harlem, where, according to recent surveys, the condition afflicts 25 percent of local children."[21]

And there is more. According to Sara Corbett writing in *Mother Jones,* an independent news magazine, "Asthma has become the most chronic illness among children in the country, affecting about 6 million kids, and its prevalence is growing. . . . According to the National Center for Health Statistics, the number of asthma sufferers in the United States has doubled since 1982.

"When an asthma patient is exposed to certain 'triggers,' . . . the lining of those airways produces extra mucus, causing them to constrict, which in turn can lead to coughing, wheezing, and difficulty breathing. The feeling is sometimes compared to breathing through a straw. A severe asthma attack can quickly starve the lungs of oxygen. According to the National Institutes of Health, more than 5,000 Americans die this way every year."[22]

An article in the *Fort Worth Weekly* in November 2005 reported that the numbers of children suffering from asthma in Texas is the highest within four specific low-income minority areas.

Although the health department was noncommittal about the causes, respiratory specialist John Fling of the University of North Texas Health Science Center was not. "Pollution in the air is playing a major role," he said, "especially the particulate matter from coal-fired power plants."[23]

Such facts and conclusions came as no surprise to Tarrant County Commissioner Roy Brooks, who told the *Fort Worth Weekly,* "There is a direct correlation between the asthma in these children . . . and the old, coal-fired power plants" of East Texas. Even though Fort Worth itself hasn't had a coal-burning plant in operation since the 1920s, "the prevailing winds blow the pollution from those plants in East Texas directly across southeast Fort Worth."

According to the same article, three of the fifty worst coal-fired power plant polluters in the nation sit about one hundred miles south and southeast of Fort Worth, and that is close enough.

And there is more. Gary Mims, writing in *The Charlotte Observer,* describes the current situation: "Asthma is America's newest childhood epidemic. . . . It is the leading medical cause of school absenteeism." That's because, "pound for pound, children generally breathe more air. As a result, their exposure to irritants is higher. In 2006 alone, at least $8 billion was spent fighting asthma."[24]

Even among those who received treatment, there were significant problems and unnecessary deaths. Another story, this time from *USA Today*:

> Phillip Hernandez was an eleven-year-old student at Lee Richmond Elementary School in Hanford, California, who suffered from severe asthma. He'd had respiratory problems since the age of three. When an asthma attack struck,

the fifth-grader would alert his teacher and then walk to the school health office to get a breathing treatment from a nebulizer there.

On May 13, 1996, he sensed an attack and headed to the office. The school nurse was busy in another school. . . .

Gasping for breath, Phillip tried to signal that he needed the nebulizer, which was stored beyond his reach, . . . [but] he collapsed before anyone could assemble the machine. "No one knew how to do it properly," [his mother, Linda Gonzalez], says. "If the school nurse had been there, she probably would have been able to help."

Phillip died shortly after.

His mother sued the school district and educational personnel in a wrongful death lawsuit. In 1999, a jury in Kings County, California, found the school negligent and awarded a $9 million judgment. The judge subsequently cut the award to $2.23 million. A California appeals court upheld that decision in 2002.[25]

According to *Salon.com,* on June 2, 2005, the same thing almost happened again to "fifteen-year-old Clare McKenna. She was gripped by an asthma attack in the middle of her class at City Honors High School in Buffalo, New York. Within seconds, McKenna, an avid volleyball and softball player, was gasping for breath. The teen collapsed as two friends helped her to the nurse's office, where her asthma medication was stored. When they finally reached the office, the door was locked, and nobody was inside."[26]

Clare could not breathe and no one could open the door. At that point the staff called 911 and also called Clare's mother.

The Bush administration budget cuts and, ironically, his No Child Left Behind program are largely responsible for the shortage of school nurses.

Most doctors say that asthma should be easy to control. Nonetheless, according to *Mother Jones* magazine, "A December 2004 study by a coalition of health groups showed that nearly a quarter of children with asthma had made at least one trip to the emergency room in [that] year and just over half had missed some amount of school or day care."[27]

If asthma attacks are largely preventable, parents are asking why more of them aren't being prevented and why several of the medications used to treat asthma are turning out to be part of the problem.

In November 2005, according to *The Washington Post,* "U.S. regulators asked the makers of three popular asthma medications to add warnings to their labels stating that the drugs could increase the chances of severe asthma episodes that could result in death.

"The warnings involve long-acting bronchodilator medicines Advair and Serevent made by GlaxoSmithKline and Foradil made by Novartis AG. Patients use them daily."[28] The *Post* says that according to the FDA, while these medications may make the attacks occur less often, they may also make them worse when they actually do occur.

At first, GlaxoSmithKline did not want to implement the warning. They claimed it would increase patients' risk. They also had a lot to lose. "Worldwide sales of Advair totaled $4.5 billion [in 2004], making it Glaxo's top-selling product. Serevent sales were $639 million, and sales of Foradil were $320 million."

On January 12, 2006, a study reported that respiratory-related deaths or life-threatening events occurred four times as often among blacks who took Serevent than among blacks who did not take the drug. Harold S. Nelson, M.D., who led the research team, said "he believes that economically disadvantaged blacks in the study may have had poorer control of their underlying disease. . . . Nelson is a professor of medicine at Denver's National

Jewish Medical and Research Center. He is also a consultant and speaker for GlaxoSmithKline, which funded the study. '[Serevent] may have relieved their symptoms but masked their worsening asthma.' "[29]

GlaxoSmithKline's shares decreased nearly 5 percent following the warning label announcement. However, the reduction in sales was expected to be offset by the introduction of a new product called Super-Advair "that was to be introduced by 2008 and was forecast to generate over $1 billion in sales by 2010." Until then, most doctors are expected to keep using Advair.[30]

The *Los Angeles Times* reported that Southern California and other parts of the country had pollution so severe that it stunted lung growth in children. "The lung damage is serious enough to lead to a lifetime of health problems and, in some cases, premature death." The eight-year study monitored 1,759 schoolchildren from fourth grade until high school graduation.[31]

"We were surprised at the magnitude of the effect we witnessed," said W. James Gauderman, one of the researchers. "It pushed a lot of kids beyond that critical threshold of low lung function. And that was a surprise."

The study, which was published in September 2004 in *The New England Journal of Medicine,* made it clear that breathing polluted air during childhood actually harms lung development.

In March 2005, a connection was reported for the first time between breathing mercury in the air from coal-fired plants and autism, a disorder with varying degrees of severity that affects normal functioning of the brain in areas of speech and communication. The study was done by scientists at the University of Texas Health Science Center in San Antonio. "This is a preliminary study," said lead author Raymond F. Palmer, an associate professor at UT–San Antonio's Department of Family and Community Medicine. "If corroborated, it would have pretty severe implications for policy."[32]

Dr. Palmer, who received a Ph.D. in preventive medicine from the University of Southern California, is actively involved in autism research. "The study looked at 254 counties and 1,200 school districts in Texas, comparing 2001 mercury emission levels with rates of autism and special education services."

The main finding, Palmer explained, was "a 17 percent increase in autism rates for every thousand pounds of mercury released."

What Palmer left unsaid is that coal-fired technology holds the possibility of saving or destroying life, of doing great good or great harm, and that politicians are given the enormous power and responsibility to use it wisely.

# 5

# THE AUTISM EPIDEMIC

## Stealing Lives Around the World

I set before you life or death, blessing or curse. Choose life then, so that you and your descendants may live.

—DEUTERONOMY 30:19

## INJECTED MERCURY AND AUTISM

AUTISM LOCKS CHILDREN INTO A WORLD OF THEIR OWN. AS THE National Autistic Society explains, its symptoms can include "a lack of speech, repetitive behaviors, little or no social interaction, withdrawal from parental and sibling contact, jerky body motions of specific limbs, head banging, hand flapping, and weird individual obsessions like eating cardboard containers or breaking certain specific objects each time the victims see them. . . . There is a four to one chance the child affected will be a boy."[1]

Many parents, doctors, and scientists now believe that the increase in autism is directly linked to the number of scheduled vaccinations babies received. During the 1990s, the vaccinations each child was mandated to have doubled from about twenty to almost forty. Many of these vaccines contained a mercury additive called thimerosal, a dangerous, highly toxic preservative that stops contamination of vaccines and preserves shelf life at very low cost to drug manufacturers.[2]

Not many Americans paid attention to autism until recently because twenty years ago only one in every 10,000 children in the

United States was affected. Now the CDC estimates that this ratio is one in every 166 children.[3]

Both the raw numbers and the ratios are shocking, especially in certain states. In Pennsylvania, for example, autism increased almost 1,600 percent in eleven years. Federal Department of Education numbers indicate that Ohio had only 22 reported cases of autism in 1992 and by 2002 the number of cases had increased to 5,146. In Illinois, over the same decade, there was an increase from 5 cases to 6,005 cases. During that decade, most states in the nation had an increase of at least 500 percent.[4]

All over America, parents reported similar stories. "A child, most often a boy who was developing socially and verbally, suddenly stopped acquiring new words and skills in the second year of life and then regressed, losing speech, cognitive abilities, and social dexterity. Children in this group are said to have regressive autism. Such late-onset autism was almost unheard of in the 1950s, '60s, and '70s. Now such cases outnumber early-onset cases five to one."[5]

According to the *Niagara Falls Reporter,* "There aren't enough special education teachers in any state to even begin to approach the problem. It costs states about $2 million for each child with autism for the first eighteen years of life. . . . Despite drug company studies from seventy years ago that concluded mercury-containing serum was not fit for cattle or dogs and all the huge, expensive federal agencies to protect us from just such a mistake—the CDC, the FDA, the National Institutes of Health, the Institute of Medicine—not one had taken the time to total up how much thimerosal and mercury had been added to the average child's intake with the new increased immunization schedule."[6]

The simple fact was that all of our children were now getting about 120 times the mercury exposure allowed by the EPA, while medical journals continued to tell parents that there wasn't a connection between thimerosal and autism.

As Robert Kennedy, Jr., points out in "Deadly Immunity," history suggests otherwise. Eli Lilly, the first pharmaceutical company to use thimerosal, "knew from the start that its products could cause damage or even death in both animals and humans. . . .

> In 1935, researchers at another vaccine manufacturer, Pittman-Moore, warned Lilly that half the dogs Pittman injected with thimerosal-based vaccines became sick, leading researchers there to declare the preservative "unsatisfactory as a serum intended for use on dogs."
>
> In the decades that followed, the evidence against thimerosal continued to mount. During the Second World War, when the Department of Defense used the preservative in vaccines on soldiers, they required Lilly to label it "poison." In 1967, a study in *Applied Microbiology* found that thimerosal killed mice when added to injected vaccines. Four years later, Lilly's own studies discerned that thimerosal was "toxic to tissue cells" in concentrations as low as 1 part per million, 100 times weaker than the concentration in a typical vaccine. Even so, the company continued to promote thimerosal as "nontoxic" and also incorporated it into topical disinfectants. In 1977, ten babies at a Toronto hospital died when an antiseptic preserved with thimerosal was dabbed onto their umbilical cords.
>
> In 1982 the FDA proposed a ban on over-the-counter products that contained thimerosal, and in 1991 the agency considered banning it from animal vaccines. But tragically, that same year the CDC recommended infants be injected with a new series of mercury-laced vaccines containing thimerosal. Newborns would be vaccinated for hepatitis B within twenty-four hours of birth, and two-month-old infants would be immunized for haemophilus influenza B and diphtheria-tetnus-pertussis."[7]

The reason was money. Thimerosal allowed drug companies to package vaccines in multiple doses that cost only half as much to produce as single doses and that was more important to them than the health risks.

The fact is that the average two-month-old baby vaccinated with thimerosal-containing shots received in a single day a dose of mercury that was between 100 and 125 times above the government's allowable exposure level. Although some children have survived this overdose without apparent damage, others, who are sensitive to mercury, have not. Children vulnerable to autism have proteins in their blood as well as immune system components that differ significantly from children without the disorder.

"For various reasons still not completely understood, the bodies of some kids are incapable of ridding themselves of mercury. In fact, an August 2003 *International Journal of Toxicology* study revealed that healthy normal children excreted eight times more mercury through their hair than did autistic children."[8]

According to Dr. Tim O'Shea, author of the book *The Sanctity of Human Blood,* "From the late 1980s on, by age two, American children have received 237 micrograms of mercury through vaccines alone."[9] The San Jose, California, medical examiner, whose articles have appeared throughout the world, explained the typical sequence. On the day of birth, "American babies received a hepatitis B mercury shot with 12 mcg of mercury, which is thirty times the safe level. At four months, they were injected with DTaP and Hib on the same day. These shots contained 50 mcg of mercury, which is sixty times the safe level. At six months, they were again vaccinated with hepatitis B and polio shots containing 62.5 mcg of mercury, which is seventy-eight times the safe level. At fifteen months they receive another 50 mcg, which is forty-one times the safe level. These one-day blasts of mercury are called 'bolus doses.' "

When the issue was finally studied, it turned out that mercury in the form of thimerosal was fifty times more toxic to susceptible infants than mercury from the consumption of fish was to young children. There are several reasons for this. "Injected mercury is far more toxic than ingested mercury. There is no blood-brain barrier in infants, so mercury accumulates in brain cells and nerves. Finally, infants under six months do not produce bile, which is necessary to excrete mercury."

Dr. O'Shea quotes Dr. Amy Holmes, who pointed out that "most mercury clears from the blood, very soon. Mercury in thimerosal is stored in the gut, liver, and brain and . . . becomes very tightly bound to the cells. Once inside those cells, or inside the blood-brain barrier, the mercury is reconverted back to its inorganic form. . . . It can then either do immediate cell damage or become latent and cause the onset of autism, brain disorders, or digestive chaos years later."

Barbara Loe Fisher's firstborn son was one of the children who had an immediate reaction. On January 11, 2001, Barbara, who is now the president and cofounder of the National Vaccine Information Center (NVIC), shared her story with the National Academy of Sciences Institute of Medicine Immunization Safety Committee.[10]

"Like most new mothers," she told the group, "I had no idea that vaccines carried any risk whatsoever."

At two and a half, her son Chris was an unusually happy and bright baby. He spoke in sentences. He loved to be with people and was starting to recognize written words in his baby books.

In 1980, on the day he was supposed to get his fourth DPT and OPV shots, he had mild diarrhea and had recently gotten over a stomach flu, but the nurse assured Barbara that it was fine to go ahead with the shots.

"When we got home," Barbara recalled in her statement, "Chris seemed quieter than usual. Several hours later, I walked into his bedroom to find him sitting in a rocking chair staring straight ahead as if he couldn't see me standing in the doorway. His face was white and his lips slightly blue, and when I called out his name, his eyes rolled back in his head, his head fell on his shoulder, and it was like he had suddenly fallen asleep sitting up. I tried but could not wake him. When I picked him up, he was like a deadweight, and I carried him to his bed, where he stayed without moving for more than six hours."

Neither Barbara nor her mother had any idea that he was having a vaccine reaction, that he belonged in an emergency room, or that health officials and the vaccine manufacturer should have been notified.

Barbara told the group that after those vaccines Chris was like a different child. He could no longer concentrate on his books or anything else for more than a few seconds. He cried easily, had diarrhea all the time, lost weight, stopped growing normally, and had chronic ear and respiratory infections. No one could explain the changes. All diagnostic tests were negative.

When Chris was six and still could not learn to read or write, he was placed in a classroom for learning-disabled children, where "he stayed throughout elementary, junior, and high school despite repeated efforts to mainstream him. Even with occupational therapy and counseling, he had a very negative educational experience, barely graduating from high school. As a young learning-disabled adult, who blessedly survived the difficult teenage years without destroying himself like some of his learning-disabled classmates, he is trying to find his place in the world, working in a mailroom and taking steps to better cope with the disabilities that made it difficult for him to learn in the classroom, so he can get more formal education."

Barbara and Chris now know that it could have been worse, that some children with vaccine reactions have "been left quadriplegic, profoundly mentally retarded, epileptic, or have died. . . . I am not alone. There are mothers whose children have suffered vaccine reactions who are being coerced by doctors to continue vaccinating with the threat that they will be reported to social services as child abusers and have their children taken from them if they don't comply. Children have been denied an education, denied health insurance by HMOs, and denied government entitlement benefits by agencies persuaded to employ a 'no vaccination, no health insurance and benefit' policy. The mothers of these children know they have a sacred duty to protect their children's lives and live in fear of state officials and even their own pediatricians."

Three years later, Barbara Loe Fisher told *Mothering Magazine* that in the spring of 1982 she had watched the Emmy Award–winning NBC-TV documentary *DPT: Vaccine Roulette,* produced by Lea Thompson. Barbara called the TV station and asked them to send her the medical research that had been used to document the show. According to Barbara,

> There, in the pages of *Pediatrics, The New England Journal of Medicine, The Lancet,* and the *British Medical Journal,* I finally found clinical descriptions of reactions to the pertussis vaccine that exactly matched the symptoms I had witnessed. . . . I was stunned. I felt betrayed by a medical profession I had revered all my life.
>
> Vaccine-induced brain injuries appear to be on a continuum ranging from milder forms such as ADD or ADHD and learning disabilities, to autism-spectrum and seizure disorders, to severe mental retardation, all the way to death. On this continuum, and often coinciding with brain dysfunction, is immune system dysfunction ranging from development of severe allergies and asthma to intestinal bowel disorders, rheumatoid arthritis, and diabetes. . . . My son's vaccine reaction nearly a quarter century ago is identical to those that thousands of other mothers have reported to the National Vaccine Information Center (NVIC) for the past twenty-two years.[11]

Barbara Loe Fisher's son, born in 1978, was part of the first wave of children suffering from autism and other disabilities now believed by many to be linked to thimerosal. Twenty-one years later, "on a Friday afternoon in July 1999 a joint statement by the the American Academy of Pediatrics and the Public Health Service was released to the press advising Americans that the amount of mercury in vaccines administered to children, through a preservative called thimerosal, exceeded federal health guidelines." This statement hid the anxiety, the fear, and the sense of panic that federal health officials had been struggling with for weeks, ever since levels of mercury had been added up by "dumbfounded" federal officials who had been responding to a broad FDA inquiry regarding mercury in consumer products. Some officials claimed this was the first time they had ever heard mercury was even in vaccines. Others obviously knew and had worried about it for years. Even so, the statement "downplayed the risk of mercury injected in newborns, and it downplayed the degree to which mercury exceeded federal safety standards. Doing the simple math, a child following the recommended schedule and receiving vaccines at birth, two months, four months, and six months was receiving mercury in excess of the EPA safe standards by a factor of 36×, 120×, 77×, and 66×, respectively. That's 120 times the safe federal standard." It was also clear that the start of the sharp increase in autism and other neurodevelopmental disorders matched the change in the vaccine schedule.[12]

Six months later, in June of 2000, fifty-two people were invited to a "secret unprecedented" emergency meeting at the Simpsonwood retreat and conference center in Norcross, Georgia. The secret meeting had been arranged by the CDC to discuss mercury in vaccines and the autism epidemic. It included the most important health officials in the field, representatives of every major vaccine manufacturer, including Merck, Wyeth, Aventis Pasteur, and GlaxoSmithKline, the top vaccine specialist for the World Health Organization, and high-level staff from the FDA and CDC.

The meeting had been arranged because, according to PutChildrenFirst.org, Dr. Tom Verstraeten, a top CDC epidemiologist, "was in a near panic." The data he was analyzing were showing a statistically significant and frightening relationship between the amount of mercury children were receiving through their vaccines and autism. "No matter how he tried to run the numbers, he wrote, the association 'just won't go away.' "[13]

According to the actual transcripts of the Simpsonwood meeting, which were stamped with the words DO NOT COPY OR RELEASE and CONFIDENTIAL but were obtained by PutChildrenFirst.org under the Freedom of Information Act, Verstraeten told everyone at the meeting that he had been stunned by what he saw. He explained that since 1991, when three additional vaccines containing thimerosal were given to extremely young infants, the estimated cases of autism had increased fifteenfold.

"Forgive this personal comment," Dr. Robert Johnson, an immunologist from the University of Colorado who attended the meeting at Simpsonwood, responded to Verstraeten's information by saying, "but I got called out at eight o'clock for an emergency call and my daughter-in-law delivered a son by C-section. Our first male in the line of the next generation, and I do not want that grandson to get a thimerosal-containing vaccine until we know better what is going on. It will probably take a long time. . . . I know there are probably implications for this internationally, but in the meanwhile, I think I want that grandson to only be given thimerosal-free vaccines."

Dr. John Clements of the World Health Organization felt differently: "The research results have to be handled," he is quoted in the same article as saying, "and even if this committee decides that there is no association and that information gets out, the work has been done, and through the Freedom of Information, that will be taken by others and will be used in other ways beyond the control of this group. . . . My mandate as I sit here in this group is to make sure at the end of the day that 100,000,000 are immunized with DTP, hepatitis B, and if possible Hib, this year, next year, and for many years to come, and that will have to be with thimerosal-containing vaccines."

According to additional documents obtained by Robert Kennedy, Jr., under the Freedom of Information Act, "The CDC paid the Institute of Medicine to conduct a new study on the risks of thimerosal." Their researchers ruled out "the chemical's link to autism. It withheld Verstraeten's findings, even though they had been slated for immediate publication and told other scientists that his original data had been 'lost' and could not be replicated. . . . By the time Verstraeten finally published his new study in 2003, he had gone to work for GlaxoSmithKline and reworked his data to bury the link between thimerosal and autism."[14]

In October of 2000, the Department of Health and Human Services (HHS) secretary, Donna Shalala, along with Congressman Dan Burton and the American Academy of Pediatrics (AAP), finally requested a "voluntary recall" of all vaccines containing thimerosal.

Just as with cap and trade mercury contamination from coal-fired plants, the people making the money and doing the damage were, once again, left in charge of making the decisions.

Congressman Burton was distressed. Perhaps it was because, as David Kirby pointed out in *Evidence of Harm,* Burton "had a grandson named Christian, who had become desperately ill after receiving multiple vaccinations against nine different childhood diseases in one day. Most of them contained thimerosal. Within days, Christian became completely unaware of his surroundings. He would run aimlessly around the house, screaming indiscriminately, flapping his arms, and banging his head against the walls. Doctors diagnosed autism. Dan Burton blamed thimerosal."[15]

"We all know and accept that mercury is a neurotoxin," Burton said. "And yet the FDA has failed to recall the fifty vaccines that contain thimerosal. Every day that mercury-containing vaccines remain on the market is another day that the Department of Health and Human Services is putting 8,000 children at risk."[16]

Nevertheless, except for the voluntary recall, that's exactly what happened. Three years later, in the November 2003 issue of *Pediatrics,* the CDC published its own study clearing thimerosal of any link to autism in children.

Then, having closed the issue in 2004, the CDC expanded the network of children endangered by thimerosal by reintroducing it into the American marketplace. This time they recommended that children six to twenty-three months old get annual flu vaccines and added that since there was "no proof of harm" from exposure to thimerosal, the CDC would not advise parents or doctors to choose a mercury-free version of the flu vaccine.

As of 2005, both FDA and CDC officials were still allowing vaccine makers to add a full dose of thimerosal to flu vaccines that were injected into pregnant women, children, and the elderly.

According to Dr. Hugh Fundenburg, one of the most quoted biologists of our time, with nearly 850 papers in peer review journals, it isn't just children who are at risk. Fundenburg writes that "if an adult receives too many consecutive flu shots containing thimerosal, his or her chance of developing Alzheimer's disease is ten times greater than if he or she had one, two, or no shots." When asked why this was so, Dr. Fundenburg stated that "the gradual mercury and aluminum buildup in the brain causes eventual cognitive dysfunction."[17]

Dr. Boyd Haley at the University of Kentucky, quoted in the same article, agreed: "Mercury dramatically reduces the viability of a major brain protein called tubulin but has little, if any, effect on another major protein called actin. Both tubulin and actin are critically important for the growth of dendrites and maintenance of the axon structures of neurons. Exposing neurons to mercury rapidly results in the stripping of tubulin from the axon structure, leaving bare neurofibrils that form the tangles that are the diagnostic hallmark of Alzheimer's disease. Thimerosal, like mercury, rapidly reduces the viability of tubulin; in addition, however, it also abolishes the viability of actin . . . and would work in concert to damage neurons of the central nervous system."

Although this is still preliminary information, the connection between thimerosal and Alzheimer's disease could be critical and deserves further study.

"The available scientific evidence has not shown thimerosal-containing vaccines to be harmful," the CDC insisted when they made the flu shot decisions. They did not mention that the mercury-free flu vaccine would cost about four dollars more per shot and would be somewhat harder to make in large quantities than the vaccine using thimerosal. The AAP, which has a membership of 57,000 physicians, then backed the CDC in its *Journal of Pediatrics*.[18]

The AAP has consistently supported the position of the vaccine manufacturers in the thimerosal controversy. Its journal has reportedly received significant funding from major drug companies.

The drug companies that are still using thimerosal have also had the support of then Senate Majority Leader Bill Frist. He received $873,000 in contributions from the pharmaceutical industry, then worked to free vaccine makers from liability in 4,200 lawsuits that had been filed by the parents of brain-injured children.[19]

One of those parents is Lyn Redwood. On the evening of November 13, 2002, Lyn got a call from a friend who was a lawyer. As David Kirby reported in *Evidence of Harm*, the House of Representatives was planning to vote on the Homeland Security Act. "The lawyer told Lyn that some unnamed agent had secretly inserted a last-minute provision into the bill, adding two brief paragraphs onto the massive document before the roll call. The provision would dismiss hundreds of civil suits filed by parents against Eli Lilly and other drug companies."[20]

The rider had been written a year earlier and incorporated into a vaccine injury bill. Its author was Bill Frist.

In response, Lyn Redwood and parents of other autistic children had created the Coalition for SafeMinds (Sensible Action for Ending Mercury-Induced Neurological Disorders).

According to Kirby, the House Committee on Government Reform's chairman, Dan Burton, demanded that the rider be removed. "These provisions don't belong in this bill," he said. "This is not a homeland security issue. This is a fairness issue."

"We have an epidemic on our hands," Burton warned in a statement released the same day. "More and more parents believe that the autism affecting their children is related to a mercury preservative used in numerous vaccines given to their children. These provisions in the Homeland Security Bill will cut off their recourse to the courts, and that's just wrong. Instead of passing legislation to take away the rights of families with vaccine-injured children, we should be passing legislation to try to help them."

Frist opposed Burton's motion. "We are a nation at risk," Frist told the Senate. "The threat of liability should not become a barrier to the protection of the American people." As Kirby points out, "The implication was that thimerosal liability protection was essential if companies were going to develop vaccines against bioterrorism weapons like smallpox and anthrax. But Frist failed to mention that these vaccines contain no thimerosal whatsoever.

"The effort to kill the rider failed, despite intense lobbying by parents and their allies on Capitol Hill. The next day, the Senate passed the House version of the bill in its entirety. That same day, Bush administration lawyers quietly filed a motion in the federal

Vaccine Court to permanently seal the records on all thimerosal-related material handed over to the government."

Congress ultimately repealed the measure in 2003, but by then Frist had inserted yet another provision into an anti-terrorism bill that would deny compensation to the families of children suffering from autism and other vaccine-related brain disorders. "The lawsuits are of such a magnitude that they could put vaccine producers out of business and limit our capacity to deal with a biological attack by terrorists," explained Dean Rosen, health policy adviser to Frist.[21]

The following year, another vaccine containing thimerosal was introduced. "Keeping your baby well this winter just got a little easier," wrote *Babytalk* magazine in October 2004. "An annual flu shot is now recommended for all healthy babies over six months of age during the October to March flu season."[22]

*Babytalk* magazine did not mention thimerosal but went on to say, "According to new guidelines from the American Academy of Pediatrics (AAP) and the CDC, this is good news since young children are at higher risk of being hospitalized for the flu. . . . Parents should bring their child in for the immunization as soon as it is available." Margaret Rennels, M.D., chairman of the committee on infectious diseases for the AAP, added, "Babies who have received the vaccine in the past need only one shot, while those who are being vaccinated against the flu for the first time will need a second one a month after the first."

In the fall of 2004, about 100 million flu vaccines were produced and only 6 to 8 million of them were available without thimerosal.

Then, on June 25, 2005, *The New York Times* ran a front-page article that quoted Dr. Marie McCormick, who chaired the Institute of Medicine's Immunization Safety Review Committee, saying, "It's really terrifying, the scientific illiteracy that supports these [autism] suspicions."[23]

According to Robert Kennedy, Jr., four years earlier, in January

2001, the same Dr. McCormick had herself told fellow researchers, "The CDC 'wants us to declare, well, that these things are pretty safe. . . . We are not ever going to come down that [autism] is a true side effect' of thimerosal exposure.

"The committee's chief staffer, Kathleen Stratton, had also predicted, in advance, that the Institute of Medicine would conclude that the evidence was 'inadequate to accept or reject a causal relation' between thimerosal and autism. That, she added, was the result 'Walt wants,' a reference to Dr. Walter Orenstein, director of the National Immunization Program for the CDC."[24]

To those on the inside, it was not surprising that in May 2005 the Institute of Medicine issued its final report saying that there was no proven link between autism and thimerosal in vaccines.

According to the *Naples Sun Times,* Steven Goodman, M.D., MHS, Ph.D., an associate professor at Johns Hopkins School of Medicine and a member of the Institute of Medicine panel, backed off from the report's certainty. He said, "It was clear from the report that we were not giving thimerosal a clean bill of health. . . . Mercury is definitely a neurotoxin. . . . We only said that the evidence favored that there was not a connection between autism and thimerosal exposure."[25]

In 2005, after a year-long fight, Dr. Mark Geier, who heads the Genetics Center of America, and his son David obtained records from the CDC. Over the next year, the two men demonstrated, through half a dozen studies, that there was an astonishing connection between autism and thimerosal. At first, they could hardly believe their own results. They reran the numbers again and again, but each time they were identical. There was no autism among the children who had never received thimerosal.

"The CDC is guilty of incompetence and gross negligence," says Mark Blaxill, vice president of SafeMinds. "The damage caused by vaccine exposure is massive. It's bigger than asbestos, bigger than tobacco, bigger than anything you've ever seen."[26]

In February of 2005, SafeMinds announced that it endorsed HR 881, introduced by Congressman Dave Weldon, M.D. (R-FL), to "assure the removal of mercury from all vaccines" in America. If passed, HR 881 would "prohibit children under the age of three and pregnant women from receiving flu vaccines that contain more than one microgram of mercury beginning in the 2006–2007 flu season." The next year, the age for children prohibited from being injected with thimerosal would be increased to six. By 2009, this cap would finally be applied to all vaccines in the United States. "Our bill sets into place a law giving vaccine manufacturers and public health officials a realistic but firm timetable for elimination of mercury in this country," said Representative Dave Weldon.[27]

At the time of this writing, in the summer of 2006, many vaccines in America still contained thimerosal, including Pediarix given to infants, Chiron's and Aventis's flu vaccines, as well as meningitis, diphtheria, and tetanus given as boosters at age eleven. Thimerosal was also still present in more than two hundred FDA-approved drugs.

Many activists and concerned parents still wondered why it would take America until 2009 to stop injecting a known neurotoxin into fragile, unformed infants and children or, for that matter, into the rest of us. To them, Weldon's proposed bill seemed like too little too late. How many more American children's lives would be needlessly stolen for the profit and convenience of pharmaceutical companies and others in the network? And why would the HR 881 bill not include discontinuing thimerosal's use in Third World countries, where it is still administered in pre-1999 quantities and still damaging untold millions of children who are depending on American medicine for desperately needed protection from illness and disease?

In China, for example, where autism had been virtually unknown before the introduction of thimerosal, by 2005 there were more than 1.8 million autistic children. Autistic disorders also appeared to be soaring in India, Argentina, Nicaragua, and other developing countries that were being supplied with thimerosal-laced vaccines from the United States.[28]

The answer is as simple as it is heartbreaking. In 2006, the annual global market for vaccines was expected to be $10 billion. The potential loss of profits by eliminating thimerosal is still more important to drug companies and some decision makers than the destruction of innocent new lives around the world.

# 6

# THE BREAST CANCER INDUSTRY

## Mammograms, Ionizing Radiation, and Conflicts of Interest

> It is the duty of Christians and of all who look to God as the Creator to protect the environment by restoring a sense of reverence for the whole of God's creation. . . . All have a moral duty to care for the environment, not only for their own good but also for the good of future generations.
>
> —POPE JOHN PAUL II

IN JANUARY 1989, JUST FOUR MONTHS AFTER HEARING THAT HER mammogram was negative, forty-two-year-old Andrea Ravinett Martin was diagnosed with a large tumor in her right breast that had spread into the lymph nodes. Like so many other women who get mammograms, Andrea had been failed by the technique that we are told is our best defense against breast cancer.

I never met Andrea Martin. I never even saw her photograph until the month before she died, and then it was as an advertisement in *The New York Times* that said: "Warning: Andrea Martin contains fifty-nine cancer-causing chemicals."

That deeply affecting advertisement was sponsored by the Environmental Working Group. Most of our bodies contain as many cancer-causing chemicals as Andrea's did, but when I saw her photograph, I didn't know that.

As the Breast Cancer Fund explained, Andrea had never worked with or near chemicals and had often wondered how her formative years had turned her into a self-described "walking toxic waste site."

When someone asked Andrea if she regretted being tested for toxic chemicals, she answered, "I was completely blown away. There were ninety-five toxins, fifty-nine of which were carcinogens. . . . I was really angry. But I believe that knowledge is power. We're starting to learn that pollution isn't only in the air, soil, and water; it's also in us. . . . We are all living in a toxic stew and [the chemical industry is], quite literally, getting away with murder."[1]

Andrea Martin grew up in Memphis and, according to the Breast Cancer Fund, she graduated Phi Beta Kappa from Tulane University in New Orleans, and earned a master's degree in French from Tufts and a law degree from the University of California Hastings College of the Law.

When Andrea's breast cancer was discovered, she was told that she had a 40 percent chance of surviving the next five years. After struggling through conflicting treatment recommendations, she chose an aggressive regimen from the limited options available to fight her stage 3 cancer: six rounds of high-dose chemotherapy, a mastectomy, six weeks of radiation, then eight more rounds of another chemotherapy protocol.

In March 1990, during a period of remission, Andrea worked with Dianne Feinstein and helped to raise over $19 million toward Feinstein's campaign for governor.

Then Andrea found a small lump in her left breast. Doctors suggested a lumpectomy, but Andrea decided to have a mastectomy. She rejoined Feinstein's campaign two weeks after the surgery.

Over the next several years, Andrea raised money to fight breast cancer and established the Breast Cancer Fund. She also joined a team of sixteen breast cancer survivors to climb Mount Aconcagua in the Argentine Andes. This climb was followed by the Breast Cancer Fund's 1998 "Climb Against the Odds" at Mount McKinley, its 2000 "Climb Against the Odds" at Mount Fuji, and its 2003 "Climb Against the Odds" at Mount Shasta.

As if that weren't enough, Andrea also led the Breast Cancer Fund in their "Obsessed with Breasts" advertising campaign to increase awareness around the world. It showed photographs of beautiful young models with Andrea's own mastectomy scars superimposed where their breasts once were.[2]

Andrea went to New York and Washington, D.C., to bring attention to the toxic pollution we all carry in our bodies. Again, people responded. Millions were moved and alerted, just as I was, by her picture in the *New York Times* ad and by the statement she made to an interviewer: "My body is a record of the environmental history of my life."

In their tribute to Andrea, the Breast Cancer Fund wrote, "It is her dream—her dream of a world without breast cancer, a world where her daughter and other women's daughters and mothers and sisters would be safe from breast cancer—that created a legacy that will live beyond her."[3]

Despite Andrea Martin's dream, the reality is that, over the last fifty years, the possibility of contracting breast cancer has almost tripled in America. "In the 1960s, a woman's lifetime risk of breast cancer was one in twenty." Northern California, where Andrea Martin died, has seen breast cancer diagnoses skyrocket even higher. In the San Francisco Bay area, for example, by 2006 a woman's chance of contracting breast cancer was one in seven.[4]

Nancy Evans and Jeanne Rizzo of the Breast Cancer Fund said that in April 1994 Bay Area citizens found that they had "some of the highest breast cancer rates in the world: 50 percent higher than most European countries and 500 percent higher than Japan." In 2005 alone, "nearly 7,000 Bay Area women were expected to be diagnosed with breast cancer: 19 new cases each day. As many as one-third of these women are mothers with children at home. Each day, two Bay Area women will die of breast cancer."[5]

## THE POLITICS OF BREAST CANCER

The obvious questions are why is this happening and what can be done to reverse the trend? Curing breast cancer means discovering its causes and preventing it. Since diagnosing and treating breast cancer is a multibillion-dollar business, the priorities are tangled. It is in the financial interest of these huge businesses to keep research funds directed toward diagnosis through mammograms, then treatment through expensive drugs, chemotherapy, and radiation—rather than toward discovering the environmental causes that might eliminate the disease.

According to the Cancer Prevention Coalition,

> National Mammography Day is sponsored primarily by the American College of Radiology and American Cancer Society (ACS), which is strongly supported by giant industries. These include General Electric, which makes mammography machines, and DuPont Corporation, which produces the film for the mammography screening. . . .
>
> The ACS has distributed its own information on breast cancer awareness to help "inform" women, but all ten questions in its informational quiz . . . are limited to issues of early detection and mammography. Not one question discusses avoidable cancer risks and breast cancer prevention.[6]

About three million women in the United States today are living with breast cancer. Two million of them have been diagnosed, and one million have the disease but do not yet know it.

Every October is designated Breast Cancer Awareness Month, and as *Sierra Magazine*'s extensive article "Cancer, Inc." explains,[7] organizers plan events around the country to "fill the void in public communication about breast cancer." For the most part, the void is filled with the mantra: "Get a mammogram."

There is no mention of environmental research. Instead the promotion kit says, "Current research is investigating the roles of obesity, hormone replacement therapy, diet, and alcohol use."

Basically, Breast Cancer Awareness Month advises women to get a mamogram to see if they already have breast cancer and indirectly suggests that if they do, it's probably their own fault. Its main sponsor is AstraZeneca, a multinational company that produces tamoxifen used in treating cancer but that also produces the carcinogen acetochlor and several other herbicides. According to *Sierra Magazine,* "Its Perry, Ohio, chemical plant is the third largest source of potential cancer-causing pollution in the United States, releasing 53,000 pounds of recognized carcinogens into the air in 1996." AstraZeneca is also the world's third largest drug concern, valued at $67 billion.

"This is a conflict of interest unparalleled in the history of American medicine," *Sierra Magazine* quoted Dr. Samuel Epstein, a professor of occupational and environmental medicine at the University of Illinois School of Public Health. "You've got a company that's a spin-off of one of the world's biggest manufacturers of carcinogenic chemicals. They've got control of breast cancer treatment. They've got control of the chemoprevention [studies], and now they have control of cancer treatment in eleven centers—which are clearly going to be prescribing the drugs they manufacture."

According to Dr. Epstein's article in *Tikkun* magazine, the American Cancer Society's twenty-two member board was created specifically to solicit corporate contributions. Since 1982, the American Cancer Society has insisted on "unequivocal" proof that a substance causes cancer in humans before taking a position on limiting it. They are "largely indifferent to cancer prevention."[8]

After analyzing the American Cancer Society's budget and programs in 1998, the *Chronicle of Philanthropy*, a watchdog organization that monitors major charities, concluded that the agency was "more interested in accumulating wealth than saving lives."[9]

High-ranking officials in the National Cancer Institute routinely accept lucrative positions in the cancer-drug industry. Such tangled financial interests explain why the cancer establishment—the medical institutions, corporations, and agencies that control cancer research, treatment, and education—continues to ignore mounting evidence that many cases of cancer are avoidable.

Though genetics and early diagnosis are stressed by the cancer industry, the truth is that "85 to 90 percent of breast cancers cannot be explained by inherited genetic predisposition."[10]

According to Breast Cancer Action, a national education and advocacy organization dedicated to inspiring and compelling changes to end the breast cancer epidemic, "the Cancer Industry consists of corporations, organizations, and agencies that diminish or mask the extent of the cancer problem, fail to protect our health, or divert attention away from the importance of finding the causes of breast cancer. . . . This includes drug companies that, in addition to profiting from cancer treatment drugs, sometimes also produce toxic chemicals that may be contributing to the high rates of cancer in this country and increasing rates throughout the world. It also includes the polluting industries that continue to release substances we know or suspect are dangerous to our health, and the public relations firms and public agencies who protect these polluters. The Cancer Industry includes organizations like the ACS that downplay the risk of cancer from pesticides and other environmental factors and which historically have refused to take a stand on environmental regulation."[11]

Each October, the cancer industry raises money through walks and other activities. The most successful of these are the Avon walks. They are managed by a profit-making company and raise up to $12 million each. More than a third of the money goes toward covering administrative and marketing expenses.

Breast Cancer Action concluded that even the limited funding directed to research by Avon is granted to large cancer centers that already have funding through the government and pharmaceutical companies.

## COSMETICS AND CANCER RISKS

While cosmetic firms like Avon compete for the sale of pink ribbons in support of a "cure," major loopholes in federal law still allow the $35 billion a year cosmetics industry to use thousands of untested, potentially carcinogenic chemicals in their products. Contrary to what is commonly assumed, the FDA does not evaluate cosmetics for safety before they are sold. The FDA Web site clearly states that "neither cosmetic products nor cosmetic ingredients are reviewed or approved before they are sold to the public. . . . Nor does the FDA require companies to do any safety testing of their own cosmetic products before marketing."

In fact, according to the Cancer Prevention Coalition, "Cosmetics are the least regulated products under the Federal Food, Drug, and Cosmetic Act (FFDCA). The National Institute of Occupational Safety and Health found that 884 of the chemicals available for use in cosmetics have been reported to the government as toxic substances . . . which have been found to cause genetic damage, biological mutations, and cancer."[12] A U.S. General Accounting Office report noted that the FDA has not committed any resources for assessing the problems with those 884 chemicals.

The Cancer Prevention Coalition warns: "Many cosmetic manufacturers lack adequate data on safety tests and have refused to disclose the results of these tests. . . . Only 3 percent of the 4,000 to 5,000 cosmetic distributors have filed reports with the government on injuries to consumers."

Yet because the skin is extremely permeable, cosmetics are easily absorbed, especially products like makeup left on for a long time.

The FDA knows that a wide range of personal care products, including shampoos, hair conditioners, cleansers, lotions, and creams, are dangerously contaminated with intense concentrations of the highly volatile 1,4 dioxane. "The Consumer Product Safety Commission concluded that 'the presence of 1,4 dioxane, even as a trace contaminant, is a cause of concern.' These avoidable risks of cancer in numerous personal care products is inexcusable, particularly as the dioxane is readily removed by a process known as 'vacuum stripping.' "[13]

A clear liquid that dissolves in water, 1,4 dioxane can be released into the air, water, and soil. We can be exposed to 1,4 dioxane when we breathe air, drink water, or eat food that is contaminated with it. Our skin may contact it when we use cosmetics, detergents, and shampoos. If 1,4 dioxane enters the lungs it can pass into the bloodstream. Once that happens, it is distributed throughout the body and is rapidly converted into other chemicals. The EPA has admitted that 1,4 dioxane is a "probable" human carcinogen.[14]

The Cancer Prevention Coalition points out that "artificial colors such as Blue 1 and Green 3 are carcinogenic. . . . Other cosmetic colors such as D&C Red 33, FD&C Yellow 5, and FD&C Yellow 6 have been shown to cause cancer . . . when applied to the skin. Some artificial coal tar colors contain heavy metal impurities including arsenic and lead, which are carcinogenic. The use of . . . hair color products, particularly black and dark brown colors, is associated with increased incidence of human cancer, including non-Hodgkin's lymphoma, multiple myeloma, and Hodgkin's disease."[15]

About 50 million American women frequently use hair dyes. Most have no idea that they are at risk. The FDA claims that it is powerless to effectively protect women against the tar dyes because of an exemption in the 1938 Food, Drug, and Cosmetic Act. That doesn't change the fact that a hair dye might be made up of up to fifteen different active ingredients, including those that "contain or produce carcinogenic contaminants. . . . The first study, in 1976, looking at breast cancer and hair dye use reported that of 100 consecutive breast cancer patients seen in a clinical practice in New York, 87 had been long-term users of hair-coloring agents. . . . Women most vulnerable were in the fifty-to-seventy-nine-year-old age group, suggesting that the latency period between the damage begun by hair dyes and the result can be lengthy."[16]

Women who have been coloring their hair for twenty-four years or more have a higher risk of developing non-Hodgkin's lymphoma. According to a study by researchers at Yale School of Medicine, "Women who dyed their hair starting before 1980 were one-third more likely to develop non-Hodgkin's lymphoma. Those who used the darkest dyes for more than twenty-five years were twice as likely to develop the cancer. . . . The Yale researchers did not find any larger risk of cancer in women who started using hair dye in 1980 or later. 'This could either reflect the change in hair dye formula contents over the past two decades, or indicate that recent users are still in their induction and latent period,' explained Yawei Zhang, who also worked on the study."[17]

Although lanolin itself is safe, the Cancer Prevention Coalition points out that "cosmetic grade lanolin can be contaminated with such carcinogenic pesticides as DDT, dieldrin, and lindane, as well as other neurotoxic pesticides. Cosmetic talc is also carcinogenic. Inhaling talc and using it in the genital area is known to be associated with increased risk of ovarian cancer."[18]

Many of these same carcinogenic chemicals are also being used in baby care products. According to the Mintel International Group, a market research firm that tracks cosmetics, "children's personal care" products are a $300 million a year business, with sales up 65 percent since 1999. People in the industry added that the "pricier kids' grooming products are the ones with the fastest-growing sales."[19]

According to another survey by researchers at Loyola University Medical Center in Maywood, Illinois, "in the first month of life alone the average newborn is bathed four times a week and exposed to forty-eight different ingredients in baby washes, shampoos, and lotions."[20] Doctors caution parents that the risk for children is increased because their skin is thinner.

"Children's skin to body mass ratio is much higher, so you also have more product absorption," explained Nanette Silverberg, M.D., director of pediatric dermatology at St. Luke's–Roosevelt Hospital Center in Manhattan, as quoted in the same article. "A child's immature liver and kidneys may also have trouble processing any chemicals that are absorbed through the skin." Experts also point out that "some common additives can irritate skin," such as "parabens, used as preservatives in many lotions and cleansers; boric acid, listed as sodium borate on labels, used as an emulsifier in some diaper rash ointments and baby skin creams; and foaming agents like sodium laurel sulfate, used in soaps and shampoos. . . . Diethanolamine, an emulsifier found in some baby oils and lotions, has also been linked to cancer in studies in laboratory animals."

While the United States waits for proof of harm, the European Union has aggressively targeted all chemicals that are believed to cause cancer, birth defects, and infertility. The European Parliament and the European Council jointly passed a rule banning hundreds of known or probable carcinogens from perfume, makeup, hair dye, and other cosmetic products, but in the United States the battle for safer cosmetics has barely gotten off the ground. Avon has been an early target.

In August 2002, a small group of activists calling themselves Follow the Money: An Alliance for Accountability in Breast Cancer (FTM) confronted Avon on how it has run its famous three-day fund-raising walks and how it distributes the substantial monies generated from them and the sale of products bearing the pink ribbon. They did not publicly question the sincerity of the company's commitment to defeat breast cancer but said that the company's position as the largest corporate contributor to breast cancer research awareness and support programs had invited a type of scrutiny that Avon seemed to wish would disappear.

FTM, which consisted of eleven state and local breast cancer groups, emphasized that to a casual observer, the Avon Foundation's Web site appeared to provide "ample information on its breast cancer grants," but efforts to analyze Avon's giving patterns were stymied by their refusal to share more information on the needs assessment process that is used to decide who will receive the grants. A list of names of the thirty-person advisory group and information about environmental research were also withheld. In addition, the information on the site accounted for the disbursement of only $145 million of the $165 million that was raised since 1993.[21]

"What we've been saying," explained Barbara Brenner, executive director of Breast Cancer Action, one of the groups that has challenged Avon, "is that since Avon has positioned itself as a champion of breast cancer [patients], it should be open to scrutiny." Barbara elaborated: "It's not just about their research grants and fund-raising walks. Everyone who bathes a baby or wants to wear cosmetics is entitled to know they are safe. You cannot depend on the company to set the standard, because they are in the business of making money. We're in the business of trying to save lives."[22]

Brenner's group had special concerns about the safety of two ingredients in the cosmetics made by Avon and many other cosmetic companies: parabens and phthalates. "Industry has told us forever that this stuff is safe," Brenner said. "They also told us that it doesn't penetrate our skin. It turns out we're swimming in this stuff. We ought to be looking at it more closely."

According to the same source,

> Parabens are derived from a petroleum base and are used as a preservative in everything from baby products, to shampoo and mascara, to deodorant. They prevent fungal and bacterial growth and give toiletries shelf lives of many months. Recent studies have shown that parabens act like an estrogen in the body. They easily penetrate the skin. Breast Cancer Action is worried that since exposure to external estrogens has been shown to increase the risk of breast cancer, repeated exposure to parabens in cosmetics might also promote the growth of cancerous cells. . . .
>
> Phthalates are a family of chemicals that are clear liquids resembling common vegetable oil. The larger-molecule phthalates make plastics flexible and can be found in everything from kids' toys to kitchen flooring. The smaller-molecule phthalates are used to make the time-release coatings on drugs. They also help make adhesives, lubricants, weather stripping, and safety glass. Four phthalates in particular (DMP, DEP, DBP, and DEHP) are used in cosmetics and personal care products. DPB gives nail polish a plasticlike consistency that makes it flexible and chip resistant. The other three phthalates are used in perfumes. When perfume

fragrances are dissolved in DEP, DMP, or DEHP, they evaporate more slowly, making the scent linger longer.

As we sat together at a long conference table in her San Francisco office, Barbara Brenner explained that she is concerned with keeping people alive and cancer free partly because of her personal experience with breast cancer. She is one of millions of smart, active professional women who suddenly, in the middle of a lucrative career and busy, productive life, discovered a lump in her breast that changed everything.

"I've had breast cancer twice," Barbara told me as she shuffled a stack of newsletters on her desk. "I was forty-one the first time. There was no breast cancer in my family. There were no risk factors except that I had not had children. I took care of myself. I ate well. I exercised regularly. I was practicing law, doing civil rights work. Actually, I was always an activist and always outspoken. Getting breast cancer just changed the focus of my activism.

"One morning, on vacation, I did my breast self-exam and discovered a lump that turned out to be cancer. I had a lumpectomy followed by chemotherapy and radiation. I was luckier than many women, because I didn't have to work and I got great care. My friends came out of the woodwork to support me. But, in the midst of all that, suddenly I didn't know how long I had left to live. That changed my perspective, so I decided to spend whatever time I had doing this work. As it happened, I had a recurrence in the same breast. Again, I discovered it doing a self-exam. Both times my mammograms didn't show anything. I learned the hard way that, with a lumpectomy, about 10 percent of cancers come back. This time, I had a mastectomy." Barbara paused, then added, "I thought if I ran fast enough, maybe it wouldn't catch me, and I'm still running.

"Cancer is not a pretty pink ribbon," she added with emotion. "People aren't sailing through their treatments. They are bald, and throwing up, and some of the treatments actually increase their risk for other cancers.

"We figured that if people knew what was really happening with the Cancer Industry, they would be furious, and that's exactly why we are telling them. We are good at channeling anger into action that makes a difference. We take a broad view. We think that if we can change what is happening in breast cancer, we can change what is happening in the whole world of environmental contamination."

I asked Barbara to explain what she meant.

She pushed her short dark hair back from her forehead and leaned forward. "The politics of poisoning of the planet are not likely to change any time soon," she said. "So, the only thing that can change is our understanding of it. Most people who are aware think about how they can personally avoid A, B, or C, and that's fine as far as it goes. But the problem is that a lot of it is unavoidable and even with those things that we can personally avoid, everybody else will still be exposed. We need to go further. We need to organize in our communities to see that safer methods for controlling dangerous chemicals are used. The answers will not come from the top. The answers will only be driven by the people who are looking for change. What we are really talking about is changing how we live, how we think about life, and how we make money. Keep in mind that the people who are profiting from this are getting more sophisticated every day in how they deceive the public and fight against public awareness. These groups use two major scare tactics. First, they argue that jobs will be lost if practices are changed. Second, they use the language of science to argue that there is not enough information to prove this or that. Instead, we need to focus on evidence of harm and how we can prevent harm in the first place. If you talk about proof in the scientific sense, there is never going to be enough proof.

"I think it would be really interesting for people in the U.S. to learn what is being done in Europe. Because, here in America, there is something that amounts to hubris. To the extent that short-term profits are valued over life, there is a tragic flaw. Capitalism is killing us. The truth is that making our products safer is really in industry's interest because the people who use the products would live longer, but nobody is taking a long-term view.

"Avon is considered a socially responsible company," Brenner continued. "They are supposed to empower women with their search for the cure. But why should a cosmetic company be making cancer research decisions? Yet it's not by accident that almost every cosmetics company is involved in breast cancer research. For them, it's really just another marketing strategy. Cosmetics firms should get out of the breast cancer field, because, with the chemicals they use, it is a conflict of interest for them. For years, Avon and others have been using dangerous untested chemicals. They are the symbol of a much larger problem. When you get past them, you find the pharmaceutical companies who are also using dangerous chemicals and also doing 'research'! But real prevention is against their financial interest. Even when you look closely at many health advocacy movements, you will find that they are funded by the very same companies that make money treating the illness they are trying to cure.

"So, we have the option to throw up our hands and give up. But I prefer not to because I know that if we don't act and act soon, what our generation is facing is going to look like a walk in the park."

Rather than throw up her hands, Brenner decided to start investigating Avon precisely because it is led by women and is the largest corporate sponsor of both the Avon walks and what she calls the other breast cancer stuff, which often passes for research:

> First, [BCA] tackled the company's use of parabens. According to Avon's own U.S.-based Web site, eighty-two of

> their products contained parabens. . . . BCA purchased one share of Avon stock and became a shareholder. It then approached three larger shareholders who were also "socially responsible": Domini Social Investments, Trillium Asset Management, and Walden Asset Management. BCA convinced them to join in sponsoring a shareholder resolution demanding that Avon study the feasibility of removing parabens completely, or substituting a nonestrogenic alternative.
>
> "Initially, we simply signed on to a statement urging Avon to sit down and talk with these breast cancer activists and listen to what they had to say," Adam Kanzer, director of shareholder advocacy for Domini, told the *Journal of the National Cancer Institute*. "We were quite surprised when we didn't get any response to that request, and then we decided to examine the issue more seriously."
>
> Kanzer explained that Avon's first response was to approach the U.S. Securities and Exchange Commission (SEC), where they attempted to prevent the Breast Cancer Action proposal from appearing before the shareholders. Avon cited the efficacy of parabens in reducing the risk of microbial contamination.[23]

Avon later said: "We believe that discontinuation of the use of parabens and replacement with an inferior preservative would present a potential health risk to our consumers that is neither necessary nor warranted."[24] With that, the board of directors convinced its other stockholders to vote against the resolution.

Breast Cancer Action decided to turn its attention away from parabens to phthalates, at least temporarily. This time they and the other activist shareholders planned to put a new, stronger resolution before the shareholders that would force Avon to study alternatives to phthalates. At that point, Avon suddenly reversed its position and made an unexpected announcement. "They said they were taking, or in some cases had already taken, phthalates out of their products."

"Removing these chemicals is a small but important step by a corporate giant," Brenner said. "It is important for all of the people Avon markets to, especially women of childbearing age, and it's important for future generations. It is not our goal to terrify people," she added. "But we really want them to understand that the only people working in their best interest are people who are not making money by selling these products."

An Avon spokesman predictably announced that the move was just part of the company's "wish to allay public concern," not an indication of a safety concern on their part.

Under increasing public pressure in January 2004, L'Oréal, Revlon, and Unilever announced that they, too, had eliminated phthalates from their cosmetic lines.

A new study in 2004 seemed to give credibility to Brenner's earlier fear that parabens, which still remained on the American market, might be implicated in breast cancer:

> Microbiologist Philippa Darbre of the University of Reading, England, found parabens in the breast cancer tumors of eighteen of the twenty women she tested. The very small study doesn't show definitively that the parabens caused the breast cancer, but according to Dr. Darbre, there was evidence that the parabens came from beauty care products.
>
> The parabens she found were the "ester-bearing form of parabens," which indicates they came from something that had been applied to the skin, such as an underarm deodorant, cream, or body spray. When parabens are eaten they are

> metabolized and lose the ester group, making them less strongly estrogen mimicking.
>
> [Dr. Darbre] suggested they might have come from deodorant. "One would expect tumors to occur evenly, with 20 percent arising in each of the five areas of the breast," she said. "But up to 60 percent of all breast tumors are found in just one-fifth of the breast—the upper outer quadrant, nearest the underarm."[25]

According to this source, the Campaign for Safe Cosmetics is still deeply concerned about parabens and other chemicals. They are asking all cosmetics companies to sign the Compact for Cosmetics, which is a pledge to immediately remove all European Union–banned chemicals and develop a plan to replace additional chemicals of concern with safe alternatives within three years. As of December 2005, that had not been done.

Barbara Brenner explained to me that while there is relatively little we can say with certainty about the causes of breast cancer, removing known or suspected carcinogens is a vital first step. "There are two causes of breast cancer we know about for sure," she added. "One is estrogen, which is what many women have been taking in hormone replacement therapy (HRT). The other is ionizing radiation, which is exactly what mammograms create."

## IONIZING RADIATION AND OTHER ENVIRONMENTAL RISKS

"For years," Brenner continued, "women were told to have mammograms and to take estrogen. Then, thirty years later, a study found that estrogen was linked to breast cancer, so now experts tell women that they should decide whether to take estrogen or not. How are women supposed to know what to decide?" she asked. "Besides, for a lot of them, it may already be too late.

"Mammograms do not prevent cancer," she explained. "At best, they find it. And, at worst, they cause it." As Brenner sees it, mammography screening is a profit-driven technology creating risks that are compounded by unreliability.

The *International Journal of Health Services* explains:

> Contrary to popular belief and assurances by the U.S. media, the National Cancer Institute (NCI) and American Cancer Society (ACS), mammography is not a technique for early diagnosis. In fact, a breast cancer has usually been present for about eight years before it can finally be detected. Furthermore, screening should be recognized as damage control rather than misleadingly as "secondary prevention."
>
> Missed cancers are particularly common in premenopausal women owing to the dense and highly glandular structure of their breasts. . . . Missed cancers are also common in postmenopausal women on estrogen replacement therapy, as about 20 percent develop breast densities that make their mammograms as difficult to read as those of premenopausal women.[26]

Roughly one-third of all cancers, including those virulent enough to double in size in about a month, are diagnosed by self-exams between annual mammograms. As a result, "Women can be lulled into a false sense of security by a supposedly negative result on an annual mammogram."[27]

As Michael Moss of *The New York Times* pointed out, the problem is often exacerbated by doctors who fail to read cancer-positive mammograms accurately:

[In 1995] Dr. Kim A. Adcock started a revolution in mammography. . . . Dr. Adcock had just become radiology chief at Kaiser Permanente Colorado. . . . When he discovered that one doctor had missed ten cancers in the space of eighteen months, he fired him. Over the next two years, Dr. Adcock fired two others who were missing more than their share of tumors. He then reassigned eight doctors who were not reading enough films to stay sharp—or for the data to show how sharp they were. "I had to assume they might be dangerous," he explains. . . .

"Every mammography program in the country should be doing something like this," said Dr. Robert A. Smith, the ACS's screening chief. . . .

A yearlong examination by *The New York Times* . . . found [that] the government has fallen far short of its pledge to ensure high-quality mammography for all. In Denver, Dr. Adcock narrowed his team down to a few specialists. By contrast, most of the 20,000 doctors in the United States reading breast X-rays are generalists with limited training and practice in mammography. Many lack the skill needed, . . . yet neither they nor their patients have the tools to find out who is good and who is not. . . .

Women have been told that mammograms can find 90 percent of breast cancer. But that figure stems from ideal conditions in research. Recent real-world sampling in two states shows that doctors are finding just over 70 percent of the cancers in women who get regular exams.

Some clinics were doing much worse. Four of the six busiest centers in a study of screening in North Carolina were averaging about 65 percent. That is, they miss one cancer for every two they find. (Not all missed cancer can be blamed on the doctor; the X-rays might be poorly taken, and many tumors are simply too hard to see.)[28]

According to Dawn Prate's article on NewsTarget.com, the National Cancer Institute now admits that the false negative rate is about 40 percent among women ages forty to forty-nine. The National Institutes of Health spokespeople also acknowledge that "mammograms miss 10 percent of malignant tumors in women over fifty." On the other hand, 70 to 80 percent of all positive mammograms are in error. They do not show any presence of cancer when a biopsy is done.[29]

Many experts believe that mammography can actually cause cancer. Prate quotes Dr. Frank Rauscher, then director of the National Cancer Institute: "In 1976, mammographic technology delivered five to ten rads (radiation-absorbed doses) per screening, as compared to one rad in current screening methods. In women between the ages of thirty-five and fifty, each rad of exposure increased the risk of breast cancer by 1 percent."

And there is much more:

> According to Russell L. Blaylock, M.D. [a highly respected neurosurgeon], one estimate is that annual radiological breast exams actually increase the risk of breast cancer by 2 percent a year. So, over ten years the risk will have increased 20 percent. . . .
>
> Dr. John W. Gofman, an authority on the health effects of ionizing radiation, estimates that 75 percent of breast cancer could be prevented by avoiding or minimizing exposure to ionizing radiation. This includes mammography, X-rays, and other medical and dental sources.
>
> Since mammographic screening was introduced, the incidence of a form of breast cancer called ductal carcinoma in situ (DCIS) has increased by 328 percent. Two hundred percent of this increase is allegedly due to mammography. In addition to harmful radiation, traditional, nondigital mammography may also help spread existing cancer cells

> due to the pressure placed on the woman's breast during the procedure. According to some health practitioners, this compression could cause existing cancer cells to metastasize from the breast tissue.
>
> Cancer research has also found a gene, called oncogene AC, that is extremely sensitive to even small doses of radiation. A significant percentage of women in the United States have this gene, which would increase their risk of mammography-induced cancer. An estimated 10,000 carriers [of this gene died] of breast cancer [in 2005] due to mammography.[30]

Since the risk of radiation is higher among young women, the National Cancer Institute now reluctantly admits that, among women under thirty-five, mammography could actually cause seventy-five cases of breast cancer for every fifteen it identifies. As a result, they have finally withdrawn recommendations for premenopausal mammography.

In an effort to get beyond diagnosing breast cancer after it occurs, *State of the Evidence: 2004,* a summary of all known environmental causes of breast cancer, was compiled by the Breast Cancer Fund and edited by its health science consultant, Nancy Evans. It pointed out that in 2004 breast cancer was expected to kill more than 40,000 American women and more than 410,000 women worldwide.[31]

The study confirmed that "ionizing radiation created by mammograms and other forms of X-ray is now the best established environmental cause of breast cancer." A growing body of evidence also implicates "nonionizing radiation," like electromagnetic fields and radio frequency radiation, as possible contributors. In addition, "compelling scientific evidence links many of the 85,000 synthetic chemicals in use today by either altering hormone function or gene expression." As with ionizing radiation, some synthetic chemicals, called mutagens, can also cause gene mutations that lead to breast cancer.

There is now broad agreement that exposure over time to estrogens is another factor. So the earlier in life a woman's menstrual cycle begins and the later it ends, the higher the risk. That is also true of estrogen administered through hormone replacement therapy and of other compounds with estrogenic activity.

There is also experimental evidence that "certain chemicals can disrupt hormone function and increase the risk. These chemicals include the insecticide heptachlor, the herbicide atrazine, and the ingredients in some sunscreens." Previously mentioned phthalates and parabens, which are used in cosmetics and other personal care products, were also named in this study.

Finally, *State of the Evidence* identified "several other carcinogenic chemicals that included the pesticide DDT and some forms of PCBs, once widely used in the manufacture of electrical equipment, carbonless paper, and other industrial and consumer products."

In the early 1990s, breast cancer activists petitioned the U.S. Congress to investigate the environmental causes of a breast cancer cluster in Long Island, where rates were 30 percent higher than in the rest of the nation. The Long Island Breast Cancer Study Project (LIBCSP) was created and was controlled by the National Cancer Institute.

The project actually consisted of more than ten studies. Altogether, $30 million was spent for the research over a ten-year period. All women in Nassau and Suffolk counties who were newly diagnosed with breast cancer during a one-year period beginning in August 1996 were eligible to participate. A comparison group of women who did not have breast cancer was randomly selected from the two counties. Altogether, 1,508 women who were newly diagnosed with in situ or invasive breast cancer and a similar number of women who did not have cancer participated.[32]

The now famous results released in August 2002 left many activists, who had waited years for the findings, incredulous. News stories about the ten-year project essentially reported that environmental pollutants did not cause breast cancer.

Researchers found no association between breast cancer and pesticide exposure or PCBs. They did identify a 50 percent elevated risk of breast cancer from polycyclic aromatic hydrocarbon (PAH), a ubiquitous toxin created by cigarette smoke, motor vehicle exhaust, and smoking or burning meat. But they downplayed the risk, calling it "modest when compared to, for example, the risk of lung cancer from smoking."

Activists pointed to several flaws. First, the pesticides and PCBs that were tested had long been banned from production and use in the United States. Second, none of the more than 85,000 chemicals that are now in daily use in this country were tested. Third, no one who lives in an industrialized society is exposed to only a single pollutant at a time. Yet that is what the Long Island study tested. Fourth, many specific chemicals known to be on Long Island and suspected as carcinogens were not studied at all. Fifth, important radiation issues were not studied either alone or in synergistic combination with chemical issues although radiation was known to be a major problem on Long Island.

Dr. Janette D. Sherman, a New York physician specializing in internal medicine and toxicology and the author of *Life's Delicate Balance: Causes and Prevention of Breast Cancer,* maintains that the Long Island study might have provided the definitive answers people were demanding if its objectivity hadn't been undermined by the project receiving financial support from the research institutions specifically tied to the polluting industries and pharmaceutical companies. The result, Sherman wrote, was that "the studies had fatal technical flaws that would inevitably produce findings that ruled out the role of environmental toxins in breast cancer."[33]

Dr. Sherman said that her concern about a conflict of interest first arose when researchers on the project from SUNY Stony Brook and Batelle were awarded a $2 billion five-year contract to operate the Brookhaven National Laboratory, which was the major known source of radioactive pollution for Long Island.

The release of radioactive materials into the air and water by the Brookhaven National Laboratory and the surrounding nuclear power plants, combined with widespread chemical contamination, were long suspected to be extremely important factors in the breast cancer epidemic. But, according to Dr. Sherman, "the Long Island study did not address the radiation issue at all and only addressed the chemical contamination issue in a very limited way."

"Feel like you're being had?" wrote Karl Grossman of *The East Hampton Star*. "One word was not mentioned in the *Newsday* or *New York Times* articles: radioactivity. That is because radioactivity was specifically not part of the study. . . . Long Island is home to the leaking Brookhaven National Laboratory," Grossman added, "and downwind from nuclear power reactors in New York, New Jersey, and Connecticut." Then he quoted a letter Dr. Sherman had written, published in *The New York Times,* which pointed out that "Long Islanders were being exposed to more than 200 radioactive substances, all demonstrated carcinogens."[34]

Grossman also quoted Alice Slater, president of the Global Resource Action Center for the Environment (GRACE), a nonprofit organization that works with research policy and grassroots communities to preserve the planet for future generations. Slater had described a meeting she had with a Columbia University official in which they discussed the Long Island study. "The Columbia administrator was very candid," Ms. Slater said. "He told us that this is a National Institutes of Health study and the NIH does not want to step on the toes of the Department of Energy." The Department of Energy owns Brookhaven National Laboratory.

"Not to look at radiation on Long Island is absolutely bizarre," Ms. Slater added. "Brookhaven Lab has been dripping plutonium and strontium 90 into our water and emitting radiation from its stacks."

Dr. Sherman put it this way: "Radioactivity was something the study wanted to stay away from, because documentation of its health impacts could upset the economic investment in nuclear technology. As a result," she added, "the [Long Island] study was performed to be inconclusive by design."[35]

In his article "Cancer Research—A Super Fraud?" Robert Ryan quotes Linus Pauling, Ph.D., a two-time Nobel Prize winner: "Everyone should know that most cancer research is largely a fraud and that the major cancer research organizations are derelict in their duties to the people who support them."

"The orthodox 'war on cancer' has failed," adds Ryan, a member of the Campaign Against Fraudulent Research.[36]

"My overall assessment is that the national cancer program must be judged a qualified failure," Ryan quotes Dr. John Bailer. Dr. Bailer spent twenty years on the staff of the National Cancer Institute and was editor of its journal. "The five-year survival statistics of the American Cancer Society are very misleading. . . . They now count things that are not cancer and, because we are able to diagnose at an earlier stage of the disease, patients falsely appear to live longer. Our whole cancer research in the past twenty years has been a total failure. More people over thirty are dying from cancer than ever before. More women with mild or benign diseases are being included in statistics and then reported as being 'cured.' When government officials point to survival figures and say they are winning the war against cancer, they are using those survival rates improperly."

Environmental carcinogens are usually downplayed while extremely costly drugs keep getting tested in experimental ways. For example, in 1992, the National Cancer Institute established a study they called "chemoprevention." More than 13,000 healthy women throughout North America were given the chemotherapy drug tamoxifen to see if it would reduce their "risk" of breast cancer. Zeneca produced the tamoxifen, and the National Cancer Institute authorized $50 million in funding.

As *Sierra Magazine* reported, "The women on tamoxifen developed 44 percent fewer breast cancers but twice as many endometrial cancers, three times as many blood clots in their lungs, and 160 percent more strokes and blood clots in their legs. Major studies in Italy and Britain found no reduction of breast cancer risk from tamoxifen. Nevertheless, in October 1998 the FDA approved tamoxifen for healthy women at 'high risk,' expanding AstraZeneca's $526 million market to 29 million more women."[37]

*Sierra Magazine* also noted that "industry's efforts to stifle evidence of the environmental links to breast cancer have now even infiltrated the medical journals." First, in 1997, *The New England Journal of Medicine* ran a book review about *Living Downstream*, by Sandra Steingraber. The author of the review said, "Steingraber was 'obsessed' by environmental pollution as the cause of cancer." It was later discovered that he was a senior official at WR Grace, the chemical company that had been responsible for contaminating wells in Woburn, Massachusetts, which formed the basis for the book and movie *A Civil Action*.

There was also an editorial by toxicologist Stephen Safe of Texas A&M University. Safe wrote about studies that examined chemicals in the blood that increased breast cancer risk. He called the studies "unconvincing" and chemophobic. As it turned out, "Safe had received research funds from the Chemical Manufacturers Association six months before his article appeared."

A groundbreaking book by Dr. Samuel S. Epstein, published in 2005, *Cancer-Gate: How to Win the Losing Cancer War,* warns that, "contrary to three decades of promises, we are losing the winnable war against cancer and that the National Cancer Institute and American Cancer Society have betrayed us." Epstein chronicles how "the National Cancer Institute and the American Cancer Society are sitting on mountains of information about avoidable environmental causes of cancer, rather than making it available to the public in a systematic and understandable way. This silence has even extended to frank suppression of information, denial of the public's right to know, and the violation of human rights."[38]

Yet, according to Epstein, "the National Cancer Institute's budget has increased thirtyfold, from $150 million to $4.6 billion," and, as of 2005, "annual revenues of the American Cancer Society had reached $800 million." Epstein added that "the more money we spend on cancer, the more cancer we get." In spite of the National Cancer Institute's and American Cancer Society's overwhelming expenditures on an ongoing series of claimed miracle cancer drugs, Epstein says, "Overall cancer mortality rates have remained essentially unchanged for more than three decades."

Janette Sherman described the struggle clearly and eloquently when she wrote, "It is clear now that winning the so-called war on cancer will not be accomplished by physicians, scientists, pharmaceutical corporations, epidemiologists, geneticists, nor by the thousands employed in various governmental agencies and universities at home and abroad. It will only be won by people who understand the connection between the loss of personal health and worldwide pollution from toxic chemicals, ionizing radiation, and endocrine altering chemicals."[39]

Shortly before her death from breast cancer, Rachel Carson had also urged all people to understand that we had more than the right to be made aware of the connections between the chemicals and the radiation introduced into our environment and the illnesses that were claiming our lives. We also had the right to be protected from them. Urging us to insist on that right was her final legacy.

Now it is our turn to reassess the current situation and summon the courage, both individually and collectively, to insist on an alternative future.

# 7

# THE NEW LUNG CANCER PANDEMIC

## Third-World Children, Contaminating Nonsmokers, and the Special Risks to Women

> The real risk in our continued pollution of the environment is not the Earth . . . but rather, it is mankind. The Earth will undoubtedly survive our depredations and will continue to swarm with life, but humankind may be extinguished and end this stage of God's experiment on the Earth.
>
> —RABBI SAUL BERMAN, YESHIVA UNIVERSITY

## THE YOUNGEST VICTIMS

WHEN I WAS TWELVE YEARS OLD, MY NEW BEST FRIEND IN THE seventh grade took out a pack of Marlboros. It was the first time she had asked me over to visit.

"Wanna drag?" she asked, lighting one up.

I shook my head.

"Go ahead," she urged. "It's cool."

I shook my head again.

"What's the matter? Are you chicken?"

Her name was Debbie. She wore boys' boot-cut Levi's and a black garrison belt. She was tall and thin and tough and beautiful.

And in that moment, in my transition from childhood to adolescence, I wanted her to like me more than anything else I could think of. I choked. She laughed. We spent the whole afternoon practicing. Little by little, she taught me how to inhale the smoke deep into my lungs. We went to town, put thirty-five cents into a cigarette machine at Jahn's, the local ice-cream parlor, and pushed the button that said "Marlboro." I hid the cigarettes deep in the bottom of my school bag so my mother wouldn't see.

The next morning at seven thirty, on the way to the school bus stop, I pulled out the Marlboros and managed to smoke a whole cigarette before the bus came.

Within days, I was hooked, which is just what the cigarette manufacturers had in mind. By the time I finally quit, I had packed in more than twenty years. I was short of breath and my lungs ached every time I inhaled, but stopping was one of the toughest things I can ever remember doing. I tried chewing gum, drinking tomato juice laced with Tabasco sauce, and just sweating it out. For months, each time I fell asleep, I dreamed I was smoking. I haven't smoked in more than two decades, but that still doesn't mean I won't get lung cancer.

For fifty years, the truth about addiction and disease from smoking was deliberately hidden from the public. Now we know that smoking is the single largest cause of premature death from chemical contamination in the United States. Now we know that millions of people suffer from smoking-related illnesses like cancer, emphysema, heart attacks, and strokes. Now we also know that at least 30 percent of cancer deaths in the United States are smoking related.[1]

Despite that certain knowledge, it is still what we don't know about the newest methods of deception that allows those deaths for profit to continue every year. Conflicts of interest in the U.S. government at the highest political levels have combined with new strategies to blur the truth about the enormous risks to nonsmokers, especially women, and to rapidly increasing chemical addiction of very young children around the world.

As public awareness in the United States expands, the attempt to addict kids in third-world countries increases. The expense of giving away free samples to innocent kids all over the world is dwarfed by the long-term economic gain through deadly addiction of yet another generation.

In Hong Kong, children as young as seven commonly become addicted to the cigarettes they receive free through industry-sponsored promotions. Global Solidarity Against Big Tobacco reports that in Cambodia, Philip Morris hires pretty young girls to give away cigarettes on the street to young boys. The girls often light the cigarettes first so that the boys will smoke them right away. In West Africa, the company has sponsored huge concerts. During the performances cigarettes are often handed to children as young as ten.[2]

In Eastern Europe, young girls also give out cigarettes at both rock concerts and discos. Anyone who accepts "a light on the spot" will be given Marlboro sunglasses. In Taipei, kids often "hang out at the 'Whisky A Go-Go disco,'" where free packs of Salems are on each table. In Buenos Aires, girls in safari gear give out cigarettes to schoolchildren during lunch breaks.

"In countries where cigarette advertising is banned or restricted, sponsoring live or televised concerts allows the companies to get around local regulations." In Taiwan, when RJR Nabisco agents arranged a concert by teen idol Hsow-Yu Chang, five empty packs of Winstons were the only accepted admission ticket, and with ten a souvenir sweatshirt was included. Sports sponsorship is another magnet because it implies that smoking and fitness mix. Young people see cigarette logos linked with their heroes and also with health, excitement, speed, and triumph. In America, cigarette companies evade the federal ban on TV advertising of tobacco and receive valuable TV airtime by sponsoring sports events.[3]

Ross Hammond and Andy Rowell offer a glimpse into how tobacco marketing works:

> Fritz Gahagan, who once worked as a marketing consultant for the tobacco industry, offered insight into how the tobacco industry has dealt with one of its most intractable issues. "The problem was, how do you sell death? How do you sell a poison that kills 350,000 people a year, 1,000 people a day? You do it with the great open spaces . . . the mountains, the lakes. . . . They do it with healthy young people. They do it with athletes. How could a whiff of a cigarette be of any harm in a situation like that? It couldn't be—there's too much fresh air, too much health, too much absolute exuding of youth and vitality. That's the way they do it."[4]

Outside the United States, the central messages are wealth, health, and consumption—in short, be like us in the USA. According to Kenyan physician Paul Wangai, "Many African children have two hopes. One is to go to heaven, the other to go to America."[5] Tobacco companies in America exploit both of those dreams by associating wealth, glamour, and freedom with smoking. It is a cruel trick. All over the world tabacco giants addict children by using slogans that appeal to their dreams of America like "L&M: The Way America Tastes," "Winston: The Spirit of the U.S.A.," and "Lucky Strikes: An American Original."

Children are given Marlboro T-shirts in Kenya and Marlboro clothing in Guatemala. In Tongo, even baby-sized Marlboro-decorated clothing is given away. In Thailand, cigarette logos have appeared on kites, T-shirts, pants, notebooks, earrings, and chewing gum packages. In Kuala Lumpur, Malaysia, a record store wraps customers' tapes in advertisements for Salem cigarettes.[6]

As a result of all this and more, smoking in the developing

world rose more than 70 percent in the last twenty-five years. At the same time, the tobacco industry insists that it does not want children to smoke.

As Global Solidarity Against Big Tobacco pointed out, "80,000 to 100,000 young people are drawn in, tricked, then newly addicted to the chemicals in tobacco every day and 4.9 million people die every year from tobacco-related illnesses. By 2030, that toll is expected to rise to over 10 million deaths each year. That is the equivalent of ninety-five jet planes crashing every single day of the year. The growth plan will keep increasing the death toll, with 70 percent ultimately occurring in developing countries. Two hundred and fifty million of the children who are alive today will end up dying from tobacco-related chemical causes. What makes this different from other deaths caused by chemical contamination is that it is not a by-product of something else. Tobacco is the only consumer product that causes disease and death when used exactly as intended by the manufacturer."[7]

Not only is there a whole industry targeting young children with full knowledge of the consequences, but cigarette manufacturers purposely use special chemical additives to make cigarettes that provide extremely high levels of "free" nicotine, in order to increase the addictive speed.

As Clive Bates, Dr. Martin Jarvis, and Dr. Gregory Connolly point out in an ASH (Action on Smoking and Health) report, the additives are also used to enhance the taste of tobacco smoke. Although that may seem harmless, it's not. The addition of flavorings is designed to make cigarettes more appealing to children and other first-time users. Eugenol and menthol numb the throat so that kids cannot feel the smoke's aggravating effects. Additives like cocoa dilate the airways, allowing easier and deeper passage into the lungs and exposing the body to more nicotine and higher levels of tar. Some of the additives are toxic or addictive in their own right or in combination. When these additives are burned, new products are formed, and these, too, may be toxic or pharmacologically active.

Nicotine, which is rapidly absorbed into the blood through the surface of the lungs, mouth, and throat, reaches the brain within ten seconds. Receptors in the brain respond to nicotine stimulation by producing dopamines and other neurotransmitters that give young people what is variously described as a hit, kick, or impact. The receptors very quickly become conditioned to expect the nicotine, and when deprived of it, people experience withdrawal. This pharmacological impact and withdrawal, enhanced by the desire to be accepted or be cool, makes smoking hard to resist.[8]

By the time people figure out what has happened to them, it is often too late. The result is frequently an early death.

In the United States alone, "the tobacco industry loses about 5,000 customers every day. About 1,200 of those die. Another 3,500 manage to stop smoking." The easiest new "replacement smokers" are children too young to understand the dangers. Sixty percent of smokers in America begin as I did, before they are fourteen years old.[9]

## THE REAL RISK TO NONSMOKERS AND WOMEN

The truth is that even people who don't smoke are placed at much more extreme risk for lung cancer and other illnesses from contaminated air than we have been led to believe, especially children and women.

A study of 1,800 nonsmoking women from five urban areas across the United States indicates that nonsmoking women increase their lung cancer risk by 24 percent when they live with a smoker. "Their risk is increased by 39 percent when they work with people who smoke and an astounding 50 percent when they hang out with smokers in such social settings as bars and restaurants."[10]

A 2005 report in *Time* magazine acknowledged that "women who don't smoke but live or work with people who do have a 27 percent increased risk of breast cancer and are as much as twice as likely to develop cervical tumors. Another study showed that children raised by smokers have as much as three times greater risk of developing lung cancer when they grow up. A fourth study found that the grandchildren of women who smoked while pregnant are more than twice as likely to develop asthma as children whose grandmothers did not. Even at low levels, secondhand smoke was associated with lower reading, math, and logic skills in children and teenagers. In 2003, Pueblo, Colorado, banned smoking in restaurants, offices, and other indoor spaces. The number of heart attacks among Pueblo residents fell 27 percent."[11]

The fact is that many extremely toxic gasses are present in even higher concentrations in sidestream smoke than in mainstream smoke and nearly 85 percent of the smoke in a room results from sidestream smoke.

"Since the 1970s, scientific evidence has accumulated proving that exposure to sidestream and mainstream environmental tobacco smoke (ETS) is a serious health hazard."[12] That evidence still continues to be largely suppressed.

In the '70s, the tobacco industry quietly began conducting "research" to counter the claims. Industry has also added more chemical additives to fool the public by masking the smell of smoke and reducing the irritating effects. The primary effort was to "modify the perception of danger," rather than actually reduce it.[13]

In 1992, the International Agency for Research on Cancer (IARC), a branch of the World Health Organization, undertook the largest European study on lung cancer caused by secondhand smoke. By September 1993, Philip Morris had established what sounded like a high-level multidisciplinary task force to undertake a similar study. Their actual plan was to "manage and monitor the public's perception of the IARC's study results by producing conflicting results."[14]

Around 1996, articles started showing up in Europe that compared the risk of lung cancer from passive smoking with "a diet high in saturated fat or drinking one or two glasses of whole milk each day."[15] The article titled "What Risks Do You Take?" made concern about secondhand smoke sound ridiculous. Such frequently appearing "news stories" were often contradicted or reversed by real reports, leading most readers to distrust all pronouncements from experts, what's called the "now they're saying" syndrome.

The tobacco industry circulated an internal document that said, "Doubt is our product since it is the best means of competing with the 'body of fact.' " Readers were asked to write for a copy of a report called "Environmental Tobacco Smoke and Lung Cancer: An Evaluation of the Risk." It was written by a team of authors referred to as "the European Working Group on Environmental Tobacco Smoke and Lung Cancer." The title of the European Working Group had the ring of authority to it, although, in fact, it was just another industry-funded enterprise.[16]

The truth is that in the United States alone secondhand smoke kills about 15,000 people every year. About 20 percent of all women with lung cancer and 10 percent of all men with lung cancer never smoked.

Dana Reeve was one of them. She was a relatively unknown singer and actress until she married Christopher Reeve. After he became paralyzed in a horseback-riding accident in 1995, Dana became widely admired as a caregiver and activist.

Americans were shocked in August 2005 when Dana was diagnosed with lung cancer. A nonsmoker all of her life, Dana was exposed to secondhand smoke in the clubs she sang at. When she announced her illness she said she was optimistic about her prognosis. Eight months later she was dead. "All over the United States, even women who had never met her grieved as if they had lost a close friend."[17]

According to *USA Today,* "The audience on ABC's *The View* gasped when Barbara Walters announced the news. In Annapolis, Maryland, Lori Ezell instant-messaged her husband the minute she heard. In Fort Wayne, Indiana, Jennifer Bosk, alternating between tears and a headache, was 'knocked off my feet.' In Birmingham, Alabama, Joyce Norman stopped in her tracks. 'I thought, "Why am I feeling this much for somebody I don't even know?"' "[18]

The deluge of condolences was "just a testament to Dana Reeve," said Maggie Goldberg, spokeswoman for the Christopher Reeve Foundation, the Short Hills, New Jersey, based organization that promotes research into spinal-cord injuries.

The same *USA Today* article quoted Dana Reeve's simple words: "I learned a long time ago that life just isn't fair, so you have to stop expecting it to be."

Lung cancer has not been fair to women.

> Deaths from the disease have increased about 150 percent over the last two decades. Between 1950 and 1997, they skyrocketed by more than 600 percent. The number of lung cancer deaths for men, however, only increased 20 percent in the last two decades.
>
> In 1950, lung cancer accounted for only 3 percent of all cancer deaths among women. By 2000, it accounted for about 25 percent of those deaths.
>
> "Lung cancer actually causes more deaths among women than breast, uterine, and ovarian cancers combined," said Dr. Jyoti Patel, an oncologist at Northwestern Memorial Hospital in Chicago and a specialist in women's lung cancer. "Women who smoked when they were young, stopped

> thirty years ago, and did all the right things since are still underestimating their risk." Patel, who coauthored an April 2004 report in the *Journal of the American Medical Association* on the risks of lung cancer in women, added that women's lack of awareness is often compounded by ill-informed doctors. "When I was in medical school," she said, "we were taught that lung cancer was a disease for men sixty years and older who smoked, and most interns today still think that is typical. As a result, many doctors don't talk to their female patients about the risks of lung cancer."[19]

The fact that so few of us, including doctors, know who the victims of lung cancer really are or how they get it is a frightening sign that the tobacco industry has once again succeeded in twisting and suppressing the truth.

While there has been very little research into the subject, there is some evidence that men's and women's hormonal differences could affect the course of the disease and also make secondhand smoke especially dangerous to women.

In 1972, Heather Crowe was a single mother living in Ottawa, Canada. Over the years, "she had worked in half a dozen restaurants, sometimes pulling three shifts a day to support herself and her daughter. She never smoked a cigarette, but after forty years of serving up eggs and coffee she was diagnosed with inoperable lung cancer in 2002. . . .

"Ms. Crowe fought successfully for full worker's compensation benefits and then became the public face of secondhand smoke risks when she appeared in . . . posters and TV campaigns. . . .

" 'I just want legislation to protect all workers [from secondhand smoke],' she said, struggling to find words at times. 'I wish this on nobody, smokers or nonsmokers.' "[20] When she was diagnosed, doctors told her she would live for about ten months. Four years later, in January 2006, Heather began having difficulty speaking and concentrating, so she checked herself into a hospital.

At that point, doctors told her that the lung tumor had spread to her brain and right arm. Along with pain medication, Heather tried acupuncture to soothe eleven painful tumors that had now spread throughout her body.

As a result of her efforts, the Smoke-Free Ontario Act, which made Ontario workplaces, bars, restaurants, and casinos smoke free, took effect shortly after her death on May 31, 2006. Canada is now considered to be among those countries at the forefront of anti-smoking legislation.

By May 2006, "ten states in the United States had also passed smoke-free workplace legislation that included offices, restaurants, bars, bingo halls, bowling alleys, nightclubs, casinos, and public transportation. These states were California, Delaware, New York, Connecticut, Maine, Massachusetts, Rhode Island, Montana, Vermont, and New Jersey.

"California passed some of the toughest and most extensive anti-smoking legislation anywhere in the world. By 2004, San Francisco prohibited smoking in all city-owned parks, shopping malls, and public sports facilities."[21]

The UK government had also outlined plans for a smoking ban in workplaces across England and Wales. Only private member clubs and pubs that did not serve food would be exempt. The health bill, first published on October 27, 2005, said the ban would be in place by the summer of 2007.

However, by April 2006 tobacco companies in the UK were fighting back. They were planning to evade the 2007 smoking ban "by giving thousands of pounds to pubs to spend on their outside spaces" so customers could continue to smoke. Pub parking lots would be converted into "gardens and ramshackle seating areas would be spruced up" and fitted with heaters as the tobacco giants channeled their advertising money into "makeovers." "Under the agreements, pubs were obliged to stock certain cigarette brands exclusively in exchange for handouts of thousands of pounds.

"Professor Gerard Hastings of the Centre for Tobacco Control Research at Stirling University said he was not surprised by the cigarette companies' strategies. He said, 'They will do all they can to keep their business going without any concern for public health.' "[22]

Despite all the bans, according to Michael Smith, an analyst at JP Morgan Chase in London, "Shares of the three biggest tobacco companies, Altria Group, British American Tobacco, and Japan Tobacco, are still surging to record levels." Profits keep rising as cigarette makers continue to expand in relatively new markets like Indonesia, Russia, Pakistan, Bangladesh, Iran, and Ukraine. At the same time, "the legal risk in the United States is declining" because more investors realize that "long-term smokers often respond to public smoking bans by smoking in different places rather than quitting."[23]

By the time these long-term smokers develop symptoms and finally quit, it is often too late. In October 2002 at the age of forty-two, Karen Blair, who'd been smoking since the age of fourteen, was diagnosed with lung cancer. The previous June, she had developed a respiratory illness. She got better, but the cough would not go away.

"I remember listening to her coughing and coughing," her brother, Neil, explained as we sat talking in his New Hampshire bed-and-breakfast. "I was lying awake while she was visiting. She had always been healthy. I made her promise to see her doctor as soon as she got back home. She followed my advice. The doctor listened to her breathing, did a chest X-ray, and found a plural effusion. That's like a blister filled with fluid. When they drained it and tested the fluid, they diagnosed the illness. But by that time it was already very advanced and very aggressive. Karen had four different protocols of chemotherapy, each with different processes and side effects. We were told that the first was very effective. It brought her cancer down to undetectable levels. We were ecstatic. We thought she was cured. But by the next month, it was as bad or worse than before. That was a huge blow.

"Karen tried three more rounds of chemotherapy, but in the meantime her lung collapsed twice and the pleural effusions continued," Neil told me sadly. "Draining them was extremely painful. Finally they put a port in her lung and went in and sprayed the lung to fuse the lining.

"After that it settled into a pattern of chemotherapy and then waiting to see the counts and find out if the chemo was working. It was a downward spiral. There never was any good news. Karen was a respiratory therapist. She worked off and on when she could, but it quickly got to the point where she couldn't work anymore. Then she couldn't walk the dog. Then she was in a wheelchair, and soon after that, she was on oxygen. Nights were especially hard for her.

"One day, she had said, 'Will you take care of my dog when I die?' Of course, I said yes. She had rescued that dog and I knew how much she loved it. After that, we planned her funeral together.

"She was a realist and, on one level, she knew what was happening, but at the same time it was always 'I'm going to beat this.'

"Sometimes, I wanted to say, 'Honey, this is too hard. Just give up.' Finally, the doctors told us that within two or three months she'd be in a coma. It turned out to be much quicker than that. She was here with me on the Tuesday before the weekend that she died. All that weekend, people close to her came. She was on morphine. During that last day and night, we had hospice care. Each time she felt like she couldn't get her breath, she'd panic. The last time, I couldn't talk her through it. She was clenching her teeth. 'Am I dying now?' she asked. I said, 'Yes, and it's OK.' We all wanted her to feel that she had our permission to stop fighting and stop suffering. I gave her the morphine. She settled into a fitful sleep, and by nine that morning, she was dead."

## CANCER TREATMENT, PROFIT, AND POLITICS

Neil and Karen learned the hard way what Dr. Ralph Moss, author of *The Cancer Industry*, pointed out:

> "Effective" cancer treatment is a matter of definition. The FDA defines an "effective" drug as one that achieves a 50 percent or more reduction in tumor size for twenty-eight days. In the vast majority of cases, there is absolutely no correlation between a tumor shrinking for twenty-eight days and the cure of the cancer or even extension of life.
>
> When the cancer patient hears the doctor say the word "effective," he or she thinks that means it cures cancer. But all it means is a temporary reduction in the size of the tumor.[24]

The treatment doesn't improve the quality of life, either. According to Dr. Moss:

> Chemotherapeutic drugs are the most toxic substances ever deliberately put into the human body. They are known poisons, they are designed poisons. This whole thing originated from experiments with "mustard gas," the horrible chemical warfare agent from World War I.
>
> Dr. John Cairns, a microbiology professor at Harvard, who published his views in *Scientific American* in 1985,

> said, "Chemotherapy was not getting very far with the vast majority of cancers."...
>
> Alan C. Nixon, Ph.D., former president of the ACS, said, "As a chemist trained to interpret data, it is incomprehensible to me that physicians can ignore the clear evidence that chemotherapy does much, much more harm than good."...
>
> "To the cancer establishment, a cancer patient is a profit center," Lee Cowden, M.D., of the University of Texas Medical School [added]. "The actual clinical and scientific evidence does not support the claims of the Cancer Industry. Conventional cancer treatments are in place as the law of the land, because they pay, not heal, the best. Decades of the politics of cancer as usual have kept us from knowing this and will continue to do so unless we wake up to this reality."

Karen was just one of 4.9 million people who died a tobacco-related lung cancer death in 2005. Karen's family, like so many other trusting Americans who hope that a cure will be found, requested that instead of sending flowers, mourners make a financial contribution to the American Cancer Society.

What they did not realize was astutely pointed out by Dr. Samuel Epstein in the *International Journal of Health Services,* where he wrote, "The American Cancer Society is fixated on damage control, diagnosis, treatment and basic molecular biology, with indifference or even hostility to cancer prevention. This myopic mindset is compounded by interlocking conflicts of interest . . . the non-profit status of the Society is in sharp conflict with its high overhead and expenses, excessive reserves of assets and contributions to political parties. All attempts to reform the Society over the past two decades have failed; a national economic boycott of the Society is long overdue."[25]

Epstein and his collegues went on to point out that "the incidence of cancer in the United States and other major industrialized nations has escalated to epidemic proportions over recent decades, and greater increases are expected." He adds that "this modern epidemic does not reflect 'a lack of resources' in the U.S. cancer establishment." The National Cancer Institute's and the American Cancer Society's budgets "have increased twentyfold since the passage of the 1971 National Cancer Act, while funding for research and public information on primary prevention remains minimal."[26]

Millions of people with lung cancer who never smoked are the tragic victims of indirect chemical contamination, but people with lung cancer who did smoke have also been victimized. First, by the tobacco industry that has knowingly addicted them. Second, by the public perception that they are to blame for their disease. Third, by the Cancer Industry, which is more concerned with high profits than with real causes or real cures. And finally, by frightening conflicts of interest and political relationships that exist between the tobacco industry, the Cancer Industry, and American politicians who take money and favors in exchange for looking the other way while smokers die by the millions.

The *Common Cause Magazine* has an interesting story:

> Former surgeon general C. Everett Koop expressed his disgust. "The support of politicians and political parties by those associated with tobacco interests is unconscionable," he said. "How can Americans believe political promises for health-care reform when both parties seem to be associated with an industry that disseminates disease, disability, and death?"[27]

Koop was right. The tobacco industry had friends throughout Congress from all parts of the country. From 1989 to 1994, for example, 73 percent of senators accepted campaign contributions from the industry. In the House, 66 percent did the same.

> In addition to political contributions, well-connected lobbyists and junkets, . . . the industry benefits from the inner politics that govern the legislative system. Tobacco control bills are often blocked . . . from only a handful have ever reached the House or Senate floor. Anti-tobacco measures have also been killed by Senate filibusters, or the threat of them, and by closed-door conference committee negotiations.
>
> "The public doesn't usually see that kind of activity," Representative Henry Waxman said. "They don't know that the tobacco interests were put ahead of the public interest." . . .
>
> "There is a political cost for challenging tobacco," admitted a former congressional aide. "You're denied an easy source of [campaign] money." Industry opponents who've . . . sworn off tobacco money often pay an even higher price when the industry finances the campaigns of their challengers. More than anything else, observers said, campaign contributions from the tobacco industry buy silence and inaction. . . .
>
> "[By 1995] Republican control had put the tobacco companies in the strongest position they had been in decades," said Representative Dick Durbin (D-IL), a leading tobacco opponent.
>
> Representative Thomas Bliley, Jr. (R-VA), who chaired the powerful House Commerce Committee, received more in PAC contributions from the tobacco industry . . . than any other Congressmember. . . .
>
> The industry's expenditure of political money goes far

> beyond campaign and soft money contributions. . . . From 1986 to 1990, . . . the tobacco industry paid members almost $1.3 million in honoraria. . . . Honoraria "payments" were easy money for members. Dick Durbin remembers a "speech" he gave to the Tobacco Institute; being new to Congress, Durbin says he "obviously didn't know how the process worked," and had spent a lot of time preparing his speech. Arriving at the group's Washington office, Durbin recalled, "I was shocked at what I found. . . . The audience consisted of the lobbyist, who gave me the [honoraria] check."
>
> The influence gained by the industry's money doesn't only come from the large amount it spends but also from the wide range of groups and activities on which it spends it. "They subsidize everything," Durbin said. "There is not a single group on Capitol Hill that you won't find them involved in."

By June 1998, it seemed clear that an important and comprehensive agreement to curb teen smoking and settle anti-tobacco lawsuits with forty states was not going to pass. Outspoken columnist Molly Ivins put it this way: "As we watched the tobacco bill die an unnatural death . . . it left only sour satisfaction for those of us who believe money runs American politics. We now have the clearest, most definitive proof. . . . Money counts more than the public interest, more than children's health and more than people's lives. . . . It was different when we thought they didn't know or weren't sure or were just ignoring the evidence. But now we know that they knew—that they have known for decades—that they were killing people. And they kept on doing it for profit."[28]

"Twenty-six thousand Texans will die this year from smoking-related illnesses," *The Nation* wrote in an extensive article published in November 1999:

> That's a fact that seems lost on Governor George W. Bush, whose presidential bid is being greatly assisted by money and manpower intimately associated with the tobacco industry. And if Bush's record in Texas is any indication, should he make it to the White House, the industry can feel certain that it will have a friend in the Oval Office.
>
> The tobacco industry is poised once again to provide the Republican Party with millions of dollars in unregulated "soft" money, offering a significant boost to the Bush 2000 drive. But Bush's tobacco ties reach far deeper than money; they go to the very heart of his campaign organization.
>
> Some of Bush's closest aides are allies of Philip Morris, including Karl Rove, . . . a close friend of Bush's since the early '70s and Bush's chief political adviser since the late '80s. [Rove] was formerly on the payroll of Philip Morris from 1991 to 1996 as a paid political intelligence operative. Former Republican National Committee chairman Haley Barbour is both an important player in [the] Bush 2000 [campaign] and arguably the number-one lobbyist for Philip Morris and the tobacco industry in Washington.[29]

President Clinton and the Justice Department began a federal lawsuit against the tobacco industry in an attempt to reclaim billions in expenditures spent to help smokers who were sick and dying. Bush intervened. According to *The Nation,* Bush said "that he hopes the 'era of big government is not replaced with the era of big lawsuits.'

"When Texas's then Attorney General [Dan] Morales and a team of lawyers forced the industry to agree to pay Texas $17 billion in damages, allowing the state to recover some of what it had spent over the years caring for sick and dying smokers, Bush actually sued the attorney general to block the payment of more than $2 billion due the private attorneys who handled the case. At the time, Morales said the Bush's lawsuit 'is about one thing and one thing only: a Republican presidential campaign and political contributions from big tobacco.' "

When Bush was elected president, he immediately nominated two men sympathetic to the tobacco industry to key posts. Former senator John Ashcroft of Missouri became attorney general, and Wisconsin governor Tommy Thompson became secretary of health and human services. Both were known as major opponents of any kind of tobacco reform. As secretary of health and human services, Thompson would "oversee agencies as diverse as the surgeon general's office, the CDC, and the Public Health Service, all of which deal with smoking prevention or the health effects of smoking."[30]

By the time Bush took office, the Justice Department lawsuit against the tobacco industry that began under Clinton was well under way. The Justice Department had already spent more than $100 million suing the tobacco industry.

Under Clinton's administration, the government had claimed that over a fifty-year period tobacco companies had engaged in a systematic campaign of fraud, material misrepresentations, half truths, deception, and lies that continued to this day, in order to minimize the health risks of smoking. The government had put on expert testimony and stated that a "comprehensive national smoking cessation program would cost $130 billion over twenty-five years." Then, just as the eight-month trial was winding to an end, the Bush administration unexpectedly reversed the $130 billion twenty-five-year demand and asked instead for only a $10 billion five-year program.[31]

The reversal appeared to be the result of political pressure by the office of Associate Attorney General Robert D. McCallum, Jr., a former partner in a law firm representing RJ Reynolds and also a personal friend and former classmate of Bush at Yale. *The Washington Post* called the reversal "insulting," then added, "But what's $120 billion between friends?"[32]

In September 2002 when Barbara Tarbox, a former model who had smoked since she was eleven, went for her annual checkup she felt fine. At forty-one she did not expect to hear that she had terminal lung cancer, which had already spread to her brain and bones. The doctor said she was not likely to live until Christmas.

Barbara's story was told in *Maclean's* magazine. "I didn't even have a cough," she told students in auditoriums across Canada, as she begged them not to smoke. "I had my yearly chest X-ray, and there was a little fine line. My doctor suggested double-checking with a CT scan. It was a Thursday I'll never forget. I sat down and the doctor said, 'Barb, I'm terribly, terribly sorry. We're transferring you to the Cross Cancer Institute. You have lung cancer.'

"I have massive, massive headaches. I'm losing my vision. I'm growing weak. I can be walking and suddenly I'll just drop. I fall. I tell my body, 'OK, stand up,' but it doesn't connect. I've lost forty-one pounds in less than two and a half months. My skin is cyanotic, which means purple-blue. My knuckles are black.

These color changes happen "very shortly before death," Barb added. "Your limbs are shutting down and your blood in your body is just going to your organs. I have deep purple veins that protrude on my legs. It's horrific.

"I'd just finished grade six when I started smoking. I was eleven years old and you know what? I was a star athlete in basketball, track and field, and I loved being athletic. [After] my first drag of a cigarette I threw up for days. I thought it was the most disgusting thing. But everybody who was the most popular was smoking. So, I thought, eventually, I'll get used to it, and that's what I did.

"I was in so much shock [when I heard the news]. Honest to goodness. I looked at my doctor and, like I told you, I said, 'I don't even have a cough.' And he said, 'Fifty percent of lung cancer patients will have zero symptoms, yet fifty percent of lung cancer patients at diagnosis are in terminal stage four. You're dying.'

"I don't want anybody else walking this path. My gosh, you know, I spent my life saying I was going to live to be one hundred and now I'm dying. Even my daughter said, 'Mommy, you've got to tell people what this has done to you.' She knows the importance. She knows what it's done to me.

"I have to say good-bye to the most incredible daughter I could ever have dreamed of and my husband of twenty years. There is no pain greater than the pain of cancer. I'm going to be dying very, very soon and that leaves my husband and daughter on their own. You can't imagine the pain."[33]

Barb Tarbox's words reached beyond schoolchildren. "Edmonton radio host Bob Stafford recalled a telephone call he received after she made one of a handful of appearances on his show: 'I had a truck driver phone me about ten days after the second interview that I had with Barb, and he was in tears. He told me that after listening to my interview, he threw a cigarette package out the window, and said he hadn't smoked in ten days at that point. He believed that Barb had actually saved his life.' "[34]

Sometimes Barb would rip off her leopard print hat when she spoke at schools to show the kids that the chemotherapy treatments had made her bald.

"I love hairstyles," she explained to the Canadian Broadcasting Corporation in January 2003. "I loved hair. Forty-one years of hair gone in ten bloody days. What's worse than my ripping off my hat and showing [the students] what cancer does?"

"By early March 2003, Barb had reached her goal of bringing her message to more than 50,000 Canadian teens. It was a milestone in her crusade, but by April the six-foot-tall woman weighed only eighty-five pounds. Barb made her last public appearance on April 17, 2003, when she spoke to a group of junior and senior high school students. By April 28, she checked herself into the hospital, unable to walk or eat on her own. She died on May 18, 2003, at the age of forty-two."[35]

Tobacco will kill a billion people like Barb Tarbox in this century. That is ten times the toll it took in the twentieth century.[36] It is death for profit, global mass murder: a chemically orchestrated, politically sanctioned crime against humanity. Except for targeting younger victims around the world and improving strategies to hide information about the special dangers to women, nothing has really changed.

# 8

# SAVING THE CHILDREN

## A Spiritual Crisis, a Theological Opportunity

> We are obliged to oppose the political empowerment of religious fatalists who view our environmental crisis as a mark of Armageddon and a glad tiding of redemption. We are obliged to support policies that tie economic development to environmental stewardship.
>
> —COALITION ON THE ENVIRONMENT AND JEWISH LIFE

### THE END-TIMERS

LUNG CANCER IS JUST ONE PART OF AN INTERGENERATIONAL tragedy. "In the United States alone, smoking during pregnancy causes 100,000 miscarriages or stillbirths every year."[1]

"Three thousand more American infants die because of smoking-related low birth weight that results in babies who are frail and vulnerable to many illnesses." In addition, babies whose mothers smoke are at "four times greater risk for sudden infant death syndrome."[2]

As I thought about these infants' deaths and about the additional impact of mandatory vaccinations that contain thimerosal, poisoned food and air, polluted water, contaminated breast milk, and the absorption of more than two hundred industrial chemicals and carcinogens before birth, I shivered. Then, seemingly from nowhere, a line from Exodus that I did not consciously remember came to me: "The sins of the fathers are visited upon the children unto the third and fourth generation." I wondered how the fathers of our society, the business executives, senators, congressmen, and even President Bush, himself a father, could justify stealing the health, the water, the air, the food, the very future of the next generation and all the generations that follow, for any amount of money.

The next day, I began to search religious texts, articles, and references for the answer, and I stumbled across the fact that many of the legislators and leaders who now create our public and environmental policy genuinely believe that there is no future for the earth or its children.

In an editorial called "There Is No Tomorrow," journalist Bill Moyers explained that "the radical religious right has succeeded in taking over [the Republican Party]—the country is not yet a theocracy," he said. "But the GOP is, and it is driving American politics, using God as a battering ram on almost every issue: crime and punishment, foreign policy, health care, taxation, energy, regulation, social services, and so on."[3]

Glenn Scherer, an author, freelance journalist, and syndicated environmental commentator whose articles have recently appeared in *Salon.com,* pointed out that "those same legislators were equally united and unswerving in their opposition to environmental protection."[4]

A Scripture-based justification for anti-environmentalism was puzzling until Scherer realized that "many Christian fundamentalists believe we are living in the End-Time, when the Son of God will return, the righteous will enter heaven, and sinners will be condemned to eternal hellfire." As Scherer explained, "They may also believe, along with millions of other Christian fundamentalists, that environmental destruction is not only to be disregarded but actually welcomed—even hastened—as a sign of the coming apocalypse."[5]

By some estimates, this group made up more than 40 percent of the U.S. Congress in 2006. Most were Republicans and, coupled with their own pro-business bias and that of their allies, this made a toxic combination. In the same article, Scherer writes:

> These politicians include some of the most powerful figures in the U.S. government, as well as key environmental decision makers: Senate Majority Leader Bill Frist (R-TN), Senate Majority Whip Mitch McConnell (R-KY), Senate Republican Conference Chair Rick Santorum (R-PA), Senate Republican Policy Committee Chair Jon Kyl (R-AZ), House Speaker Dennis Hastert (R-IL), House Majority Whip Roy Blunt (R-MO), U.S. Attorney General John Ashcroft, and quite possibly President Bush. . . .
>
> A 2002 *Time*/CNN poll found that 59 percent of all Americans believe that the prophecies found in the Book of Revelation are going to come true. Nearly one-quarter think the Bible predicted the 9/11 attacks. . . . In the 2000 election, the Christian right cast at least 15 million votes, or about 30 percent of those that propelled Bush into the presidency. . . . Because of its power as a voting bloc, the Christian right had the ear of much of the nation's leadership. Some of the leaders were End-Time believers themselves. Others are not. [It is an important distinction, because not everyone who believes that the Bible should be taken literally welcomes the destruction of the environment.] Either way, their votes were heavily swayed by an electoral base that eagerly awaited the looming apocalypse. And that, in turn, is devastating for those who hope for the protection of the earth, not its destruction.

Ever since the dawn of Christianity, small groups of believers have searched the Scriptures for signs of the End-

> Time and the Second Coming. Today, most of the roughly 50 million right-wing fundamentalist Christians in the United States believe in some form of End-Time theology.

Its outline, as Moyers explains, is simple: "Once Israel has occupied the rest of its 'biblical lands,' legions of the Antichrist will attack it, triggering a final showdown in the valley of Armageddon. As the Jews who have not been converted are burned, the Messiah will return for the Rapture. True believers will be lifted out of their clothes and transported to heaven, where, seated next to the right hand of God, they will watch their political and religious opponents suffer plagues of boils, sores, locusts, and frogs during the several years of tribulation that follow."[7]

Given that belief, it's not surprising, as Michael Luo of *The New York Times* wrote, that "word spread quickly in some conservative Christian circles when Israeli troops captured the Old City of Jerusalem from Arab forces in June 1967 [that] this was it: Jesus was coming.

"But Jesus did not return that day, and the world did not end with the culmination of that Arab-Israeli war. Neither did it end in 1260, when Joachim of Fiore, an influential twelth-century Italian monk, calculated it would, nor in February 1420 as predicted by the Taborites of Bohemia, nor in 1988, forty years after the formation of Israel, nor after the September 11, 2001, terrorist attacks. But it is not surprising that after the 2005 earthquake in Pakistan, the terrible flooding that followed Hurricane Katrina, and the tsunami in Asia talk of the end of the world began again."[8]

As Scherer notes, it is difficult to see how the belief that we are at End-Time could actually fuel anti-environmentalism and be used as an excuse to collect wealth and exercise personal power without considering the role of two of the born-again leaders who were also lawmakers: former House Majority Leader Tom DeLay (R-TX) and former Senate Committee on Environment and Public

Works Chair James Inhofe (R-OK). "Neither DeLay nor Inhofe include environmental protection in 'the Lord's work.'"[9] They expressed no impulse to care for the earth or its creatures, human or otherwise. Instead, "both have ranted against the EPA, calling it 'the Gestapo.'" James Inhofe wanted to remake America as a Christian state, and Tom DeLay fought to gut the Clean Air Act and Endangered Species Act.

According to Scherer, "DeLay has said bluntly that he intends to smite the 'socialist' worldview of 'secular humanists' who, he argues, control the U.S. political system, media, public schools, and universities. He called the 2000 presidential election an apocalyptic 'battle for souls,' a fight to the death against the forces of liberalism, feminism, and environmentalism that are corrupting America. The utopian dreams of such movements were doomed, argued the majority leader, because they did not stem from God. . . . This may explain why DeLay's Capitol office furnishings included a wall poster that read: 'This could be the day'—meaning Judgment Day."[10]

On September 8, 2005, a grand jury indicted Tom DeLay for criminally conspiring to use illegal corporate contributions to favor the election of more Republicans. "DeLay denounced the charges as a 'sham' and an act of 'political retribution,' perpetuated by his opponents. 'I have done nothing wrong. . . . I have violated no law, no regulation, no rule of the House,' he said." However, on April 3, 2006, DeLay announced he would not run for reelection. He added that he could serve "the conservative cause best by forming a lobbying firm that would work to support conservative issues."[11]

As Peter Perl reported in a series of personal interviews for *The Washington Post,* DeLay was "reveling in a dream come true of controlling not only the White House and the Congress but also the nation's agenda." At that time DeLay kept two leather bullwhips and a stone copy of the Ten Commandments in his office.

"He fought hard to hold down the minimum wage, repeal clean air and clean water legislation, scrap occupational health measures, protect corporations from consumer lawsuits, and otherwise liberate corporate America from government."[12]

Looking back, Peter Perl recalled DeLay as a somewhat "pathetic figure." According to Perl, DeLay started his professional life as a pest exterminator in Houston.

When first elected to Congress in 1985, DeLay told people he "would stay out all night drinking till the bars closed." He swore off hard liquor after he was "reborn" as a Christian. DeLay dated his Christian rebirth to his alcohol-hazed first year in Congress.

In the next twenty years, he came to develop a single-minded vision of how America should be. DeLay's America would acknowledge that the Constitution was inspired by the Bible; it would promote prayer and worship and would stop gun control, outlaw abortion, limit the rights of gays, curb contraception, end the constitutional separation of church and state, and adopt the Ten Commandments as guiding principles for public schools.

After his resignation, DeLay said he had prayed long and hard before God made clear that he no longer wanted DeLay to represent Texas's Twenty-second Congressional District. Instead, DeLay said, his God wanted him to be a messenger . . . on a much broader scale.

James Inhofe is also disturbing to people who care for creation. According to Scherer, he described global warming as "a hoax. . . . [He] makes major policy decisions based on heavy corporate and theological influences, flawed science, and, probably, an apocalyptic worldview—and he chairs the Senate Environment and Public Works Committee."[13]

> "I trust God with my legislative goals and the issues that are most important to my constituents," Inhofe has told

*Pentecostal Evangel* magazine. "I don't believe there is a single issue we deal with in government that hasn't been dealt with in the Scriptures." But Inhofe stayed silent about which biblical passages he applies to the environment, and he remained so when asked if End-Time beliefs influence his leadership of the most powerful environmental committee in the country.[14]

End-Timers' views are also shaped by literal interpretation of Genesis 1:28: "Fill the earth and subdue it. Have dominion over the fish of the seas, the birds of the air and all the living things that move on this earth." End-Timers assert that this passage proves that "man" is superior to nature and gives him the go-ahead to continue with "unchecked population growth and unrestrained resource use," whatever the cost.[15]

## CARING FOR THE EARTH

These End-Time beliefs are clearly not mainstream. According to *The Washington Post,* there is now growing evidence that many conservative church leaders strongly disagree with DeLay's and Inhofe's attitudes toward environmental issues:

> A growing number of evangelicals view stewardship of the environment as a responsibility mandated by God in the Bible.
>
> "The environment is a values issue," said the Reverend Ted Haggard, [then] president of the 30-million-member National Association of Evangelicals [shortly before his own disgrace over charges of homosexuality and drug use in November of 2006]. "It should be a banner issue for the Christian right."

In October [2005], the association's leaders adopted an "Evangelical Call to Civic Responsibility" that, for the first time, emphasized every Christian's duty to care for the planet and the role of government in safeguarding a sustainable environment.

"We affirm that God-given dominion is a sacred responsibility to steward the earth and not a license to abuse the creation of which we are a part," said the statement, which has been distributed to 50,000 member churches. "Because clean air, pure water, and adequate resources are crucial to public health and civic order, government has an obligation to protect its citizens from the effects of environmental degradation."

Signatories included highly visible opinion-swaying evangelical leaders such as Haggard, James Dobson of Focus on the Family, and Chuck Colson of Prison Fellowship Ministries.[16]

In the fall of 2005, *Christianity Today,* an influential evangelical magazine, said, "Christians should make it clear to governments and businesses that we are willing to adapt our lifestyles and support steps toward changes that protect the environment."[17]

The idea that man controls nature has only recently been armed with dangerous technical skill, vast political power, and the ability to control public environmental policy. Whatever their differences, most of the world's religious scholars have traditionally viewed man's role in God's creation with deep humility.

Acton Institute for the Study of Religion & Liberty has pointed out what many Christian denominations believe:

The whole creation is God's handiwork and belongs to God (Psalms 24:1). . . . In the ability God has given us

> to make choices also lies inherent danger. We can choose to disobey, to be irresponsible, to disrupt and disturb the peaceable relationship of creature and creation. We can choose to use nature's resources only for what we perceive is our own immediate interest. Such action is sin. . . . Often we tend to think of sin in terms of individual actions. Yet decisions and actions which we make as groups, communities, and societies constitute corporate sin. . . .
>
> Today the human race faces an unprecedented challenge to rediscover the role of steward in a time of extraordinary peril and promise. . . . The depletion of nonrenewable resources . . . the pollution of air, land, and water, the waste of precious materials, and the general assault on God's creation, springing from greed, arrogance, and ignorance, present the possibility of irreversible damage to the intricate natural systems upon which life depends. . . . The danger is real and great. Churches and individual Christians must take responsibility to God and neighbor seriously and respond (Ephesians 2:10).
>
> Ironically, science and technology have multiplied many times the ecological threat. The very instruments that brought great blessings and still hold much promise now threaten to bring disaster unless they are used in concert rather than in conflict with the created order.[18]

Rabbi Saul Berman, an associate professor of Jewish studies at the Stern College of Yeshiva University, has spoken of stewardship in Judaism. He explains that "the entire framework for Judaism's teachings on the environment emerged from the dynamic tension between Genesis 1:28, when God said, 'replenish the earth and subdue it; have dominion over . . . every living thing,' and Genesis 2:15, when God takes the newly created human 'and placed him in the garden of Eden to cultivate it and to guard it.' "[19] Rabbi Berman also points out that the second verse imposes stewardship upon man:

> If we love humanity, then we must act now to save it. . . . The real cause of environmental pollution, the real reason that people have brought the earth to its knees begging for relief, has nothing to do with people's excessive observance of the command of "subdue it!" . . . The real cause of abuse is the human failure to heed religious teachings against the exclusive importance of material goals. The real cause of our destruction of the environment is our total preoccupation with wealth and comfort. To the extent that science and technology have become the handmaidens of profit instead of truth, they have become part of the problem and need now to be redirected to being part of the solution.
>
> The longer-term solution to environmental problems depends upon our ability to reeducate ourselves and our children towards humility . . . and moderation. . . . Only such a reorientation, in which material excess is replaced with deep spiritual awareness of the ultimate partnership between humanity and the earth in the achievement of God's goals, can lay the foundation for a new and more healthy relationship between us and our environment.

During the last decade, more and more Jewish people have become involved in environmental issues. "Judaism teaches that, above and beyond everything, we have a responsibility to protect life, not only when we know for sure it's at risk, but when it may be at risk," said Mark Jacobs, director of the Coalition on the Environment and Jewish Life, a national coordinating body for the Jewish environmental movement. "Our work is creating an opportunity for people who care about the environment and are Jewish to exercise environmental commitment through a Jewish framework."[20]

In 1996, Thich Tri Quang, a Buddhist monk who was also the chief editor of *Giac Ngo,* a Buddhist magazine in Vietnam, spoke for Buddhists of all traditions when he said: "The awareness of protecting life and the living environment has been generated in recent time. However, in Buddhism, it is one of the main basic laws which was set out by the Buddha some twenty-five centuries ago for his students to follow. . . . The Buddha . . . saw that all beings in the universe were equal in nature, and in this phenomenal world, lives of all human and animals were inter-related, mutually developing, and inseparable. However, men have seen themselves as the smartest species of all beings. They have misused and abused their power and selfishly destroyed . . . the living environment. . . . The external environment is seriously polluted because the internal environment of the mind is seriously damaged. . . . Bottomless greed has pushed mankind to satisfy excessive and unnecessary demands. . . . By all means, they try to maximize their profits, without being concerned with the negative impact of . . . discharging toxins into the air, water, and earth, leading to environmental pollution and destruction of the ecological balance. . . . But I think it is still not too late for all religions, all strata of the society, and all nations to come together and jointly participate in the protection of the environment for all living species."[21]

On December 8, 1989, Pope John Paul II expressed the growing concern of Roman Catholics: "Faced with the widespread destruction of the environment, people everywhere are coming to understand that we cannot continue to use the goods of the earth as we have in the past. . . . A new ecological awareness is beginning to emerge. . . . The ecological crisis is a moral issue."[22]

On November 14, 1991, at the U.S. Catholic Conference, U.S.

Catholic bishops added their voice to the pope's. They said in part, "Our mistreatment of the natural world diminishes our own dignity and sacredness, not only because we are destroying resources that future generations of humans [will] need, but because we are engaging in actions that contradict what it means to be human. Our tradition calls us to protect the life and dignity of the human person, and it is increasingly clear that this task cannot be separated from the care and defense of all of creation. . . . In the face of these challenges, a new spirit of responsibility for the earth has begun to grow. . . . American Catholics are an integral part of this new awareness and action. In many small ways, we are learning more, caring more, and doing more about the environment and the threats to it. . . . We are not gods, but stewards of the earth. . . . These are matters of powerful urgency and major consequence. They constitute an exceptional call to conversion. As individuals, as institutions, as a people, we need a change of heart to preserve and protect the planet for our children and for generations yet unborn."[23]

At the Maryland Presbyterian Church in Towson, just north of Baltimore, weekly meetings are held by a group of members to care for the woods surrounding their church and also to discuss possible methods for restoring the Chesapeake Bay:

> "We share a conviction that it is our responsibility to care for the earth," said Bill Breakey, chair of the church environmental stewardship committee. "It's a God-given treasure, and we are a part of it."
>
> To put those convictions into practice, church members formed a study group to explore their spiritual relationship to the earth and discuss sustainable living. . . .
>
> Vincent Leggett, executive director of Blacks for the Chesapeake [added that] African-American churches, especially in urban areas, have always been involved with envi-

ronmental issues that affect their quality of life. . . . "We need to shatter the myth that people of color are not interested in environmental issues that affect their communities. . . . If you ask, do people want good water or clean air, the answer is yes every time. That's the essence of environmental issues, though they get dressed up in a lot of ways."[24]

## A NEW OPPORTUNITY

Through all of these statements of separate faiths and hundreds more there runs a common theme, and in that theme there lies a great opportunity that can bind us together and make us stronger. Once we are aware of our interdependent need, the wisest religious leaders of all faiths can grow even wiser. They can launch a shared global initiative, an interfaith movement in which God's will and blessing means more than any of the differences between faiths, a movement that allows dominion over the earth to remain in hands far wiser than ours. Though spiritual at its core, such a movement would also include practical elements like reframing our perspectives, our legislation, and our legal decisions.

A chemical barrage has been hurled against the earth and its people. The illnesses and crises we face as a result are too urgent for religious leaders to ignore.

There are some people who will continue to argue that the issues are too complex or that we need to do more research before we take any specific action. They will assert that regulatory decisions require scientific "proof" and say that we can only decide to limit a particular poison when we know everything about it. These are the arguments that have stalled us, again and again, year after year, and kept us from taking action. The intentional creation of doubt, the strategy of scientific "uncertainty," has allowed the toxic chemical perchlorate to invade our waterways and contaminate our food. Doubt is the mechanism that has allowed thimerosal to be injected into the world's children from birth onward. Uncertainty is the reckless path that has permitted tobacco companies to kill a 100 million people directly and poison the air in smokers' paths, leading to an unrecorded number of additional deaths. The same argument has allowed all forms of cancer, birth defects, cluster illnesses, and asthma to spread unchecked. It has brought us to where we are today.

Alternatively, we can answer that we do know enough to take action. We know that we have tampered with forces too complex for us, and we know that we have filled the world with thousands of poisons. We know that if we commit to working with great effort and cooperation, we can still undo much of the damage.

Real change—social, political, or environmental—requires committed leaders and the broadly "diverse and ceaseless efforts of great multitudes of people working at a variety of issues" but reaching for a single vision.[25] The act that can unite us and reverse our current course is an act of faith: the collective voices of religious leaders joining together, along with the congregations and individuals who stand behind and support them.

On a practical level, all it takes to start such a movement is the conviction that we all must care for the earth and its people together and the collective will to force the political changes that ban nonessential known carcinogens and impose regulations to minimize exposure to other potential carcinogens, as well as the common sense to test all new chemicals before they are dispersed among us.

As Rachel Carson pointed out, "It would be unrealistic to suppose that all chemical carcinogens could be eliminated from the modern world," but with shared determination the burden could be significantly lightened and the threat to every one of us could be greatly reduced. "The most critical effort should be made to eliminate those carcinogens that now contaminate our food, our water supplies, and our atmosphere" as well as our everyday products. These chemicals create the most dangerous and unavoidable kind of threat: "minute exposures repeated over and over throughout the years."[26]

I believe we are at a historic turning point at which such cooperation not only is entirely possible but also has already begun to happen. The National Religious Partnership for the Environment founded in 1992 was among the first to encourage and support the collective efforts of concerned people across the religious spectrum. Under their leadership, Jewish, Catholic, Eastern Orthodox, and Mainline and Evangelical Protestant members have shown how the diversity of traditions can work together for a common purpose. Its constituent groups now serve about 100 million Americans. These new-style environmentalists first gained attention in the early months of 1995 when they successfully fought to preserve the Endangered Species Act after it came under attack from GOP lawmakers. They have also focused on the threat of global warming.

When then EPA administrator Christie Whitman made the case to President Bush that the administration should take action on global warming, she cited the growing power and passion of the Partnership. "For the first time, the world's religious communities have started to engage in this issue," Whitman wrote in a memorandum to the president. "Their solutions vary widely, but the fervor of the focus is clear."[27]

In January 2005, an important new interfaith statement was signed by more than one thousand clergy in thirty-five states who were members of the National Religious Partnership for the Environment. It was called "God's Mandate: Care for Creation." The statement directly addressed George Bush's environmental policies and insisted that "there was no mandate, no majority and no 'values' message in the past election of George Bush or the Congress allowing them to rollback and oppose programs that care for God's creation."

According to Paul Gorman, executive director of the Partnership, the religious leaders came up with the idea of making a strong statement on the environment specifically to warn the White House and Republicans in Congress that "there are limits to the support they can expect from the religious community."[28] "People of faith are deeply and religiously alarmed," Gorman was quoted by the same source.

The interfaith statement also included the following passage:

> We come from communities that hold traditional values of neighborly love and respect for life. We benefit daily from laws that safeguard habitat and public health. We are proud of our nation's long-standing commitment to conservation.
>
> From these perspectives, we feel called to express great dismay and alarm at plans by the administration and the leadership of the 109th Congress to reverse and obstruct programs to protect God's creation in our land and across the planet. There are now specific proposals before the government that would jeopardize public health, clean air and water, sustainable sources of energy, safety of natural habitats, and the earth's climate, which embraces us all.

The religious leaders expressed specific concern about Bush's attempt to replace the Clean Air Act created in 1970 with the Clear Skies Initiative, which would have turned back thirty years of anti–air pollution effort.

The Clean Air Act required power plants to reduce mercury, sulfur, and nitrogen emissions by 2010. Bush's proposed Clear Skies Initiative "would allow an extra decade for the more than 400 grandfathered power plants built before 1977 to be fitted with pollution controls. It would also permit five times more mercury emissions, one and a half times more sulfur dioxide emissions, and hundreds of thousands more tons of the smog-forming nitrogen oxides. Ironically, the administration presented this legislation as the most historic cleanup in history. The truth was that simply enforcing the Clean Air Act would cut as many as five times more emissions at least a decade sooner."[29]

According to an analysis using the EPA's own methods and assumptions, the Clear Skies Initiative time lines would result in "2 million additional asthma attacks and 100,000 more premature deaths between 2005 and 2020. The costs to the public from loss of workdays, hospitalizations, emergency room visits, and loss of life would total $34 billion."[30]

Religious leaders rightly feared that additional soot, smog, and mercury from power plants would continue to increase the deaths and illnesses each year among infants and children. These leaders recognized that children were at special risk from power plant pollution, because they breathe more rapidly than adults and have more lung surface areas for their body size. They understood that pound for pound, children take in 50 percent more air than adults do and have more frequent adverse health effects from pollution. More than 25 million children already lived in counties that had violated national air quality standards and 35 million children lived within thirty miles of one or more power plants—a distance within which local communities would be directly affected by pollution from the plants. Two million of these children were already asthmatic. Tens of thousands of U.S. schools were located near outdated power plants that were already allowed to operate under "grandfather" clauses. Largely as a result of the National Religious Partnership for the Environment's efforts, Bush's Clear Skies Initiative was tied in a nine-to-nine vote in committee. It was an important first step.

According to the National Resource Defense Council's Web

site, however, there were still more than four hundred other major environmental rollbacks that had been promoted by the Bush administration. Many had already been quietly passed. Still others were expected to be introduced and voted on in the near future.

In the month following the release of "God's Mandate," another group of theologians, convened by the National Council of Churches, gathered at the National Cathedral in Washington, D.C., to work on a theological statement opposing political arguments that said the environment was not an issue that should concern Christians. The National Council of Churches released an open letter calling on Christians to reject teachings that suggest humans are "called" to exploit the earth without care for how their behavior affects the rest of God's creation. The National Council of Churches' statement, "God's Earth Is Sacred: An Open Letter to Church and Society in the United States," points out that there is both an environmental and a theological crisis: "We have listened to a false gospel that . . . continues to capture its adherents among emboldened political leaders and policy makers. . . . We have abused and exploited the earth and [its] people . . . jeopardizing the earth's capacity to sustain life as we know and love it."[31]

The National Council of Churches also identified eight norms to guide people on a new environmental path: justice, sustainability, bio-responsibility, humility, generosity, frugality, solidarity, and compassion.

Fr. Chris Bender, an Orthodox priest, referred directly to the political End-Timers when he said, "Some people say that the environment doesn't matter because the Second Coming of Christ will usher in the end of the world as we know it. To make such a statement is the height of arrogance." Then he added, "We don't know when the Lord is coming back, but we do know that one day we will have to give an account for making the environment unlivable for those who come after us, and for those who are the poorest among us."[32]

Three years earlier, Ecumenical Patriarch Bartholomew, the spiritual leader of the world's 200 million Orthodox Christians, had joined Pope John Paul II in proclaiming a moral and spiritual duty to protect the environment. The patriarch eloquently voiced his own beliefs: "To commit a crime against the natural world is a sin. . . . We witness death approaching on account of trespassing against limits that God placed on our proper use of creation." His words made headlines around the world.[33]

Then, in an unprecedented joint initiative, these heads of the long warring Catholic and Orthodox churches, Pope John Paul II and Patriarch Bartholomew, signed a declaration warning that "the world faces an environmental and social crisis. They asserted that Christians have a particular responsibility to combat it and said that they welcomed the rise of environmental consciousness that could lay the foundations for developing a join 'environmental ethos.'"[34]

It is time, past time, for people of all faiths to come together and expand their efforts to stop the reckless but powerful few among us who poison for profit and kill for greed. Children being chemically contaminated today, children with leukemia, birth defects, and autism, cannot wait for politicians who claim they are searching for "proof," then bury the proof that they find.

Our proper path is clear. Not only do the End-Timers' attitudes devalue the ten thousand generations of people who came before us, but those attitudes will also ultimately make the earth uninhabitable. It took hundreds of millions of years to create the world we live in and only about a century to contaminate its air, earth, water, food, and ever-increasing numbers of its people, especially its children.

I speak now as a journalist and a mother from no particular religious perspective. I am the granddaughter of an Orthodox rabbi and the mother of a son who was baptized into the Greek Orthodox Church, then left to pursue Buddhism, because he sought compassion above all else. I have respectfully attended Jewish, Greek Orthodox, Catholic, Episcopal, Quaker, Buddhist, Hindi, and Muslim services, but I belong to no specific church. The shared opportunity and obligation I speak of stems from my deep belief that "environment" is a modern word for God's creation and that it should be treated with care. I know that we cannot poison the earth without poisoning its people. We must take the correct path, because the wrong one will lead us to an apocalypse of our own making.

Glenn Scherer explained our current trajectory with a simple parable. He said, "Many years ago, a friend of mine introduced me to his religious grandparents who, whenever they were asked about the future, proclaimed, 'Armageddon's coming!' And they believed it. Christ was due back any day, so they never bothered to paint or shingle their house. What was the point? Over the years, I drove by their place and watched the protective layers of paint peel, the bare clapboards weather, the sills and roof rot. Eventually, the house fell into ruin and had to be torn down, leaving my friend's grandparents destitute.

"In a way, their prediction had proven right. But this humble apocalypse, a house divided against itself, was no work of God, but of man. This is a parable for the 231 Christian right–backed legislators of the 108th Congress. Their constituencies' cherished beliefs may lead to the most dangerous and destructive self-fulfilling prophecy of all time."[35]

As I looked away from my desk, my grandson Jonathan's photograph caught my eye. He was gazing skyward and his eyes were wide with trust. Jonathan, who like every baby alive today already carries more than two hundred untested industrial chemicals, pollutants, and carcinogens in his tiny body; Jonathan at eight months, delicate, vulnerable, but also miraculously strong and resilient, poised on all fours, rocking back and forth on the living room rug, scuttling backward, inching forward, falling, then rising to try again, literally born to live, to crawl, to walk, to explore, and to embrace the world. Then I wondered if he and all the others of his generation and those that follow him would have their chance. I shivered involuntarily and picked up my pen.

Yes, we stand at a dangerous crossroad. But we still have the capacity to think, to see, to feel, to understand, to experience outrage, and to act on it collectively. The choice, after all, is still ours.

This is the time for action and outrage. The efforts of every parent, every person of faith, and every person of goodwill must continue and increase. We must all become pivotal in shaping new doctrine. Religious groups and leaders must increase the strength of their power, their numbers, and their voices, because the earth has become a battlefield that is chaotic, unmanageable, and impossible to control without their wisdom and collective leadership.

It is time for the politicians, businessmen, scientists, and watchdog agencies at the highest levels, who are "playing God" or yielding to bribes and pressure, to be restrained.

It is time for the military to stop telling us its job is to protect us while it poisons our water, food, and air and sickens and kills our children.

It is time for the drug companies to stop injecting the innocent children of the world with thimerosal, a known neurotoxin, and lying about it simply because it allows them to produce vaccines in bulk and increase their profit.

It is time for the cancer industry to stop misleading women with deadly financial conflicts of interest, ionizing radiation, and false representations of the effectiveness and safety of mammograms. It is time the cancer industry began the real search for the environmental causes of cancer.

It is time for the tobacco industry and U.S. politicians to stop making deals that target women, children, and innocent bystanders for death, in exchange for votes and dollars.

It is time for every single one of us to fight as if our lives and our children's lives depend upon it—because they do!

While many of us and our children are still fortunate enough not to have reached our chemical threshold for illness, it is now clear that even in the most remote areas more and more infants, young children, and families will not escape. Victims already live in every town and village and city across the globe. They are rich and poor, young and old, newly born, stillborn, and still unborn. This book told only a few of their stories and described only a handful of more than 85,000 untested chemicals and an untold number of chemically caused illnesses.

For those of us whose lives have not yet been touched by disease and certainly for pregnant mothers whose children are not yet born, control, prevention, and awareness are imperative. Even if we cannot pin down the precise culprit that triggers a specific response, we do know that the more toxic exposure, the greater the danger.

I still believe that there is hope, even a bright side, and an opportunity that awaits us if we collectively and passionately choose to take it. I still believe that through the mobilization of good people, with and without religious affiliation, there is still enough time and power to alter our current course.

We must not throw up our hands, and we cannot remain indifferent. Why should we accept, as if there were no alternative, the casual and careless and ever-increasing contamination of the world we have been given simply for profit? Why should we allow our lives and the lives and health of our children, born and yet unborn, to be compromised and stolen by a ruthless few? Surely we have the strength and wisdom not to be mesmerized and controlled and led like animals to our own slaughter.

The human equivalent of global warming is descending upon us. It is clear that we cannot continue to poison the earth's water, food, soil, and air without also poisoning more and more of its people. I believe that buried in crisis, especially urgent crisis, there is often opportunity. I believe that with the help of the world's religious leaders and other great thinkers, we can still create a new spiritual movement that is powerful enough to bind us together with the blessings of the earth's creator and guide us toward a safer, saner, healthier, more reverent future, a future based on compassion, cooperation, and comprehension that we are dealing with life in all its possibilities and frailty.

I believe that if we choose this path, it is still possible for the people of the world to be reunited by a single leap of faith, a desire for survival, and a sense of common purpose. Finally, as a mother, I believe because I must believe . . . that it is still possible to save the children.

# EPILOGUE

## A Leap of Faith: What You Can Do, Now and Later

> Let us respond, with one unified voice: "We will do and we will hearken."
>
> —EXODUS 24:7

GIVEN THE BROAD RANGE OF OBSTACLES AND DANGERS WE ALL face, I decided it would be useful to include here some resources and Web sites that offer basic practical information from reliable sources for readers who feel they need to do something active to protect themselves and their families. For those who are willing to go further, I have included Web sites of both religious and nonreligious groups who are already working individually and collectively to care for the Earth and its children.

### PERCHLORATE RELEASES BY LOCALITY

The EPA has recently put together a highly technical chart of perchlorate releases in the United States. Although dated September 2004, as of the fall of 2006 it was still the most current information available. The site claims to represent all known perchlorate releases in America as of that date. It also categorizes contamination by intensity, by city, and by county. It lists Department of Defense sites, federal agencies, private corporate sites, utility companies, and water companies that have made significant perchlorate releases.

An unidentified employee at Senator Dianne Feinstein's Washington office who told me about the government site said, "It lists all the perchlorate releases that the public is allowed to know about." The Web site can be accessed at www.epa.gov/fedfac/documents/perchlorate.htm.

## MOST CONTAMINATED AND LEAST CONTAMINATED FOODS

For readers concerned about perchlorate and other chemicals that have seeped into our food supply, the Environmental Working Group has done excellent research demonstrating that people can reduce their pesticide exposure by 90 percent by avoiding the top-twelve most contaminated fruits and vegetables and eating the least contaminated instead. They point out that eating the twelve most contaminated fruits and vegetables will expose a person to nearly twenty pesticides a day, while eating the twelve least contaminated will expose a person to barely two pesticides a day. Their guide provides people with an excellent way to make choices that will instantly lower pesticide exposure from the food they eat.

### TWELVE MOST CONTAMINATED FRUITS AND VEGETABLES

- Apples
- Bell peppers
- Celery
- Cherries
- Imported grapes
- Nectarines
- Peaches
- Pears
- Potatoes
- Red raspberries
- Spinach
- Strawberries

TWELVE LEAST CONTAMINATED FRUITS AND VEGETABLES

- Asparagus
- Avocados
- Bananas
- Broccoli
- Cauliflower
- Corn (sweet)
- Kiwi
- Mangoes
- Onions
- Papaya
- Pineapples
- Peas (sweet)

The Working Group recommends we continue to eat plenty of fruits and vegetables, buying organic food whenever possible, especially if we are dealing with the most contaminated dozen. The Working Group also makes it clear that washing will not change the contamination of the fruits and vegetables since all tested foods were washed and prepared for normal consumption prior to testing. For more detailed information on contamination in our fruits and vegetables and what you can do about it, visit the group's Web site at www.ewg.org.

## NONTOXIC PRODUCTS FOR THE HOME

To minimize danger from the use of toxic chemicals in our homes, the Tennessee Valley Authority Regional Waste Management Department has studied and reported on nontoxic household products. Gary A. Davis and E. M. Turner at the University of Tennessee point out that many extremely toxic chemicals in the home can be eliminated simply by making the right choices at the supermarket. They make clear that until World War II and the "chemical age" that followed, most people used a limited number of simple substances to keep the house clean, odor-free, and pest-free. Simple cosmetic preparations made from eggs, oil, clay, vinegar, and herbs kept hair and skin lustrous. The garden was fertilized and pests were kept down with naturally occurring substances. Weeds were removed by hand and chrysanthemums controlled a wide spectrum of pests.

The authors of the study point out that while we do not need to return to the ways of the past to avoid exposure to house toxins, we can learn some lessons from it. For example, they suggest that we use soap-based garden insecticides instead of chemically based ones. They also list several natural products that are safe substitutes to use in the kitchen and bathroom:

**Baking Soda:** Can be used as a deodorizer in the refrigerator, on smelly carpets, on upholstery, and on vinyl. It can help deodorize drains and can also be used to clean and polish aluminum, chrome, jewelry, plastic, porcelain, silver, stainless steel, and tin. It can soften fabrics and remove certain stains. In addition, baking soda can be used as an underarm deodorant and as a toothpaste, too.

**Borax:** A naturally occurring mineral, soluble in water. It can inhibit the growth of mildew and mold, boost the cleaning power of soap or detergent, remove stains, and can be used with attractants such as sugar to kill cockroaches.

**Cornstarch:** Can be used to clean windows, polish furniture, shampoo carpets and rugs, and starch clothes.

**Lemon Juice:** Another natural deodorant that can be used to clean glass and remove stains from aluminum, clothes, and porcelain.

Tennessee Valley Authority Regional Waste Management Department offers a booklet, available online, that is filled with many pages of such tips. Their suggestions cover safe, nontoxic alternatives for nearly every real need around the home and can be found on the Web at es.eps.gov/techinfo/facts/safe-fs.html.

## PERSONAL-CARE PRODUCTS

The cosmetic industry markets thousands of personal-care products with ingredients that have not been assessed for safety by either industry or government health experts. Instead, marketing decisions are made by an industry-funded panel, which have only screened 11 percent of the 10,500 commonly used ingredients for safety. The average adult is currently exposed to over a hundred untested chemicals in personal-care products every single day.

Through a new, interactive product-safety guide called Skin Deep, produced by the Environmental Working Group, consumers can now quickly see the results of real safety tests for more than 14,000 specific personal-care products.

Readers can use the Skin Deep Web site to check for contamination first, then create their own shopping lists of products that are free of fragrances and carcinogens.

The excellent Skin Deep database is available at www.ewg.org/reports/skindeep/. The group's work on personal-care product safety is available at www.ewg.org/issues/siteindex/issues.php?issueid=5005.

## RELIGIOUS ENVIRONMENTAL GROUPS

Readers interested in obtaining a list of religious groups that work toward caring for the environment can do so by contacting the National Religious Partnership for the Environment at 49 South Pleasant Street, Suite 301, Amherst, MA 01002. The partnership can be reached by telephone at 413-253-1515. Their Web site is at www.nrpe.org.

The National Religious Partnership for the Environment is an extraordinary group that was founded in 1993 by four major religious organizations and alliances. Working together they now serve tens of millions of Americans. The group includes:

- The U.S. Conference of Catholic Bishops, comprising the bishops of the United States and the Virgin Islands
- The National Council of Churches of Christ (NCCC), a federation of thirty-four Protestant, Orthodox, and Anglican denominations
- The Coalition on Environment and Jewish Life (COEJL), an alliance of agencies and organizations across all four Jewish movements
- The Evangelical Environmental Network (EEN), a coalition of twenty-three evangelical Christian programs and educational institutions

The National Religious Partnership office can help readers find contacts in all fifty states from the Jewish, Catholic, mainline Protestant, Eastern Orthodox, and evangelical Protestant communities. A few additional resources include:

- The Center for Respect of Life and Environment: www.crle.org/index.asp
- Earth Sangha: Buddhism in Service of the Earth: www.earthsangha.org
- Indigenous Environmental Network: www.ienearth.org/iensub.html
- The Centre for Indigenous Environmental Resources (CIER): www.cier.ca
- First Nations Environmental Network: www.fnen.org
- Native Americans and the Environment: www.cier.ca
- Islamic Foundation for Ecology and Environmental Sciences: www.ifees.org/index.htm

- Orthodox Church and the Environment: www.ecupatriarchate.org/visit/html/environment.html
- Quaker Earthcare Witness: www.quakerearthcare.org
- Unitarian Universalist Ministry for Earth: www.uuaspp.org/index.shtml

## SECULAR ENVIRONMENTAL GROUPS

For readers who prefer to link up with an environmental group unaffiliated with a particular religious denomination, there are thousands all over the world. A partial list of some of the major organizations is supplied by the Natural Resources Defense Council and includes the following:

- The Alliance for Justice, a national association of environmental, civil rights, mental health, women's, children's, and consumer advocacy organizations
- The American Council for an Energy-Efficient Economy (ACEEE), dedicated to advancing energy efficiency as a means of promoting both economic prosperity and environmental protection
- The Center for a Livable Future, promoting policies to protect health, the environment, and sustainable living
- The Clean Water Network, an alliance of over a thousand organizations working to protect our nation's water resources

For their complete list, visit the Natural Resource Defense Council's Web site at www.nrdc.org.

For additional environmental groups around the globe, I suggest the World Directory of Environmental Organizations Online at www.interenvironment.org/wd. This is a descriptive directory that lists over 350 Web pages with thousands of entries and links. It has detailed subject and geographic sections, background pages, and numerous cross-references. It is actually an expanded online version of a standard reference book that has been produced since 1973 by InterEnvironment, a program from the nonprofit California Institute of Public Affairs in cooperation with the World Conservation Union and the Sierra Club.

In addition to the advice and resources listed in these sources and Web sites, I urge concerned people to carefully consider all of the information about thimerosal before vaccinating their children; to test their tap water or drink bottled water whenever possible; to avoid fish that are high in mercury as well as synthetic estrogens, smoking, environmental smoke, and all unnecessary ionizing radiation in the form of X-rays, CT scans, and mammograms.

When choosing a new home, I encourage people to consider such environmental background issues as air quality, which often goes well beyond close proximity to coal-fired plants. Previous toxic manufacturing or a former military base at the site are also important factors to consider.

Having said these things, I must stress that all of them provide only a small and limited refuge in an increasingly dangerous and uncontrollable chemical environment. What the public has been told is safe today may turn out to cause asthma, autism, birth defects, cluster illnesses, cancer, or other illnesses tomorrow. Our exposures start before we are born. They are cumulative, uncontrolled, multiple, and usually impossible to track accurately because they are affected by the time and degree of exposure, other chemical exposures, and our individual genetic predispositions.

Not being able to prove that a specific chemical has caused or been the catalyst of an unexplained illness or cluster of illnesses does not exclude the possibility. Those industries or groups who claim that their specific kind and dose of poison is safe may be telling the truth as it is understood today, and yet, when that toxic chemical is combined with other chemical exposures, it may still be causing illness. All of us are living on a steady cumulative diet of weak poisons. We need to disregard the false assurances by those chemical companies and political leaders whose main interest is making money, whatever the human cost.

# NOTES

## Foreword | Stolen Lives: A Personal Awakening


1. Manuel Roig-Franzia and Catharine Skipp, "Tainted Water in the Land of Semper Fi," *The Washington Post,* January 28, 2004.
2. Rachel Carson, *Silent Spring,* introduction, pages xv–xix, New York: Mariner Books, 2002.
3. Carson, *Silent Spring,* page 8.
4. Carson, *Silent Spring,* pages 237–38.


## 1 | Global Contamination: No One Is Spared


1. Marla Cone, "Ancestral Diet Gone Toxic," *Los Angeles Times,* January 13, 2004.
2. Cone, "Ancestral Diet Gone Toxic."
3. DeNeen L. Brown, "Toxic-tainted Arctic Animals Passing Poisons on to Inuit," *The Washington Post,* May 22, 2001.
4. Bruce E. Johansen, "The Trashing of the Arctic," *The Progressive,* December 2000.

## 2 | Deadly Water: Cluster Illnesses, Leukemia, and Birth Defects

1. Manuel Roig-Franzia and Catharine Skipp, "Tainted Water in the Land of Semper Fi," *The Washington Post,* January 28, 2004.
2. Roig-Franzia and Skipp, "Tainted Water in the Land of Semper Fi."
3. Roig-Franzia and Skipp, "Tainted Water in the Land of Semper Fi."
4. Public health assessment, U.S. Marine Corps, Camp Lejuene, Onslow County, North Carolina, pages 1–2, Agency for Toxic Substances and Disease Registry.
5. Peter Eisler, "Pollution Cleanups Pit Pentagon Against Regulators," *USA Today,* October, 14, 2004.
6. Bob Feldman, "War on Earth," *Dollars and Sense,* March/April 2003. Also see the Military Toxics Project, www.miltoxproj.org. The 2006 information is from Molly Ivins, "Credit Where Credit Is Due," *Wilmington Star-News,* June 22, 2006.
7. Comments from a field hearing by Senator Harry Reid (D-NV) of the Committee on Environment and Public Works, April 12, 2001.
8. Leukemia Research Fund, 43 Great Ormond Street, London, England WC1N 3JJ.
9. The Aplastic Anaemia Trust, www.theaat.org.uk.
10. Jeffrey St. Clair, "Fallon's Fallen," *CounterPunch,* August 7, 2002.
11. St. Clair, "Fallon's Fallen."
12. Frank X. Mullen Jr., "Scientists Say Tungsten a Promising Clue to Leukemia Cluster," *Reno Gazette-Journal,* February 6, 2003.
13. Renee Downing, "Cancer Wars," *Tucson Weekly,* February 12, 2004.
14. Downing, "Cancer Wars."
15. Craig Steinmaus, Meng Lu, Trandall L. Todd, and Allan H. Smith, "Probability Estimates for the Unique Childhood Leukemia Cluster in Fallon, Nevada, and Risks near Other U.S. Military Aviation Facilities," *Environmental Health Perspectives,* May 2004.
16. Downing, "Cancer Wars."
17. Downing, "Cancer Wars."
18. Rachel Carson, *Silent Spring,* page 42, New York: Mariner Books, 2002.
19. "Christian Ecological Imperative: A Pastoral Letter from the Social Affairs Commission, Canadian Conference of Catholic Bishops," *Western Catholic Reporter,* October 13, 2003.

20. The information regarding perchlorate contamination by state is reprinted from the official EPA Web site (www.epa.gov/fedfac/documents/perchlo rate.htm).
21. Andrew Bridges, "Rocket Fuel Pollution Strains Water Supplies; Prompts Health Fears," Associated Press, January 5, 2003.
22. Bridges, "Rocket Fuel Pollution Strains Water Supplies."
23. "Community Groups Ready to File Suit over LANL Clean Water Act Violations," UC Nuclear Free online archive, May 23, 2006.
24. Peter Waldman, "Perchlorate Runoff Flows to Water Supply of Millions: A Fuel of Cold War Defenses Ignites Health Controversy," *The Wall Street Journal,* December 16, 2002.
25. Heather Hansen, "Wrestling with a New Threat to Water Quality," New West network (www.newwest.net), March 30, 2005.
26. Bridges, "Rocket Fuel Pollution Strains Water Supplies."
27. Office of Senator Dianne Feinstein, "Perchlorates: Report on Widespread Rocket Fuel Pollution in Nation's Food and Water," Organic Consumers Association newsletter, January, 11, 2006.
28. Waldman, "Perchlorate Runoff Flows to Water Supply of Millions."
29. Waldman, "Perchlorate Runoff Flows to Water Supply of Millions."
30. Office of Senator Dianne Feinstein, "Senator Feinstein 'Disappointed' that Defense Department Missed Perchlorate Deadline," feinstein.senate.gov, May 3, 2004.
31. Waldman, "Perchlorate Runoff Flows to Water Supply of Millions."
32. Waldman, "Perchlorate Runoff Flows to Water Supply of Millions."
33. Erik Olson, Jennifer Sass, and Elliott Negin, "White House and Pentagon Bias National Academy Perchlorate Report," Natural Resources Defense Council (www.nrdc.org/media), January 10, 2005.
34. Erica Werner, "Drinking Water Debate Gets More Heated," Associated Press, January 11, 2005.
35. Office of Senator Dianne Feinstein, "21st Century Timeline of U.S. Rocket Fuel Pollution Scandal," Organic Consumers Association newsletter, January, 11, 2006.
36. Erik Olson, Jennifer Sass, and Elliott Negin, "White House, Pentagon, Industry Secretly Colluded to Skew National Academy of Sciences Perchlorate Report, Documents Show," Natural Resources Defense Council, January 10, 2005.
37. Glen Martin, "Rocket Fuel Found in Milk in California: Not Clear if Amount Imperils Children," *San Francisco Chronicle,* June 22, 2004.
38. Sarah Ruby, "Perchlorate Work Begins," *The Pinnacle,* April 25, 2003.

39. Ruby, "Perchlorate Work Begins."
40. Sarah Ruby, "Olin Knew Perchlorate Harmful," *The Pinnacle*, April 18, 2003.

## 3 | Toxic Food: Thyroid Disease, Cancer, and Pesticide Victims

1. Sarah Ruby, "State Milk Has Perchlorate," *The Pinnacle*, June 25, 2004.
2. Peter Waldman, "EPA Bans Staff from Discussing Issue of Perchlorate Pollution," *The Wall Street Journal*, April 28, 2003.
3. Ruby, "State Milk Has Perchlorate."
4. Ruby, "State Milk Has Perchlorate."
5. "Rocket Fuel Chemical Found in Organic Milk," Associated Press, MSNBC.com, December 1, 2004.
6. Sarah Ruby, "Growers Fret Over Perchlorate in Produce," *The Pinnacle*, May 30, 2003.
7. Ruby, "Growers Fret Over Perchlorate in Produce."
8. Marla Cone, "Rocket-Fuel Chemical Found in Breast Milk," *Los Angeles Times*, February 23, 2005.
9. Staff article, "Scientists Find High Levels of the Toxic Rocket Fuel Chemical in Human Breast Milk," Environmental Working Group (www.ewg.org/issues/perchlorate), February 22, 2005.
10. Cone, "Rocket-Fuel Chemical Found in Breast Milk."
11. "Scientists Find High Levels of the Toxic Rocket Fuel Chemical in Human Breast Milk."
12. Sandra Steingraber, "Tune of the Tuna Fish: When a Favorite Food Turns Toxic," *Orion*, January/February 2006.
13. Colleen Diskin, "Slew of Pollutants Found in Babies," *The North Jersey Record*, July 14, 2005.
14. "Toxic Chemicals by the Hundred Found in Blood of Newborns," Environmental News Service (www.ens-newswire.com), July 14, 2005.
15. Rita Beamish, "U.S. Farmers Use Pesticides Despite Treaty," Associated Press, November 28, 2005.
16. Kristin Collins, "Birth Defects Could Be Linked to Ag-Mark Pesticide Violations," *The News & Observer*, February 26, 2006.
17. Staff article, "Methyl Bromide in California Air," Environmental Working Group (www.ewg.org), 2006.

18. John Lantigua, "Why Was Carlitos Born This Way?" *The Palm Beach Post,* March 16, 2005.
19. Charles Rabin, "Infant Has No Limbs: Parents Blame Insecticides," *The Miami Herald,* March 2, 2006.
20. John Lantigua, "Farmworkers Sue Grower Over Baby's Birth Defects," *The Palm Beach Post,* March 2, 2006.
21. Beamish, "U.S. Farmers Use Pesticide Despite Treaty."

## 4| The Cycle of Airborne Poison: Mercury Contamination and Asthma

1. Sandra Steingraber, "Tune of the Tuna Fish: When a Favorite Food Turns Toxic," *Orion,* January/February 2006.
2. Steingraber, "Tune of the Tuna Fish."
3. Jane Kay, "U.S. Urges Limits on Eating Albacore: Concerns About Mercury Levels in Some Canned Tuna," *San Francisco Chronicle,* March 20, 2004.
4. Michael Greger, M.D., "Mercury Contamination in Fish," *Vegetarian Baby and Child,* November 11, 2003.
5. Kay, "U.S. Urges Limits on Eating Albacore."
6. Michael Bender, "Canned Tuna Mercury Riskier Than Previously Suspected," Mercury Policy Project, June 19, 2003.
7. Peter Waldman, "Mercury and Tuna: U.S. Advice Leaves Lots of Questions," *The Wall Street Journal,* August 1, 2005.
8. Roddy Scheer, "Pregnant Moms Should Avoid All Tuna," Consumers Union (consumersunion.org), June 13, 2006.
9. Waldman, "Mercury and Tuna."
10. Jane Kay, "Rich Folks Eating Fish Feed on Mercury Too," *San Francisco Chronicle,* November 5, 2002.
11. Waldman, "Mercury and Tuna."
12. Steingraber, "Tune of the Tuna Fish."
13. Melanie Warner, "With Sales Plummeting, Tuna Strikes Back," *The New York Times,* August 19, 2005.
14. Warner, "With Sales Plummeting, Tuna Strikes Back."
15. Felicity Barringer, "EPA Accused of a Predetermined Finding on Mercury," *The New York Times,* February 4, 2005.

16. "Scientists Take Aim at White House," Associated Press, MSNBC.com., February 20, 2005.
17. Julie Cart, "U.S. Scientists Say They Are Told to Alter Findings," *Los Angeles Times,* February 10, 2005.
18. Shankar Vedantam, "New EPA Mercury Rule Omits Conflicting Data," *The Washington Post,* March 22, 2005.
19. Robert Kennedy, Jr., "For the Sake of Our Children," *EarthLight* magazine, Winter 2005.
20. Richard Perez-Pena, "Study Finds Asthma in 25% of Children in Central Harlem," *The New York Times,* April 18, 2003.
21. Solana Pyne, "Coal Dirt Cheap," *The Village Voice,* September 3–9, 2003.
22. Sara Corbett, "The Asthma Trap," *Mother Jones,* March/April, 2005.
23. Betty Brink, "Ill Winds," *Fort Worth Weekly,* November 23, 2005.
24. Gary Mims, "More Preschoolers Getting Asthma," *The Charlotte Observer,* April 9, 2006.
25. Kevin McCoy, "School Blamed in Boy's Death," *USA Today,* December 13, 2005.
26. Laurie Udesky, "No School Nurses Left Behind," *Salon.com,* September 29, 2005.
27. Corbett, "The Asthma Trap."
28. "Warnings for Asthma Drugs Sought," Reuters, *The Washington Post,* November 19, 2005.
29. Salynn Boyles, "Asthma Drug May Be Deadlier for Blacks," *WebMD Health News,* January 12, 2006.
30. PBR staff writer, "GlaxoSmithKline Winded by FDA Label Clampdown," *Pharmaceutical Business Review,* November 24, 2005.
31. Miguel Bustillo, "Smog Harms Children's Lungs for Life, Study Finds," *Los Angeles Times,* September 9, 2004.
32. Todd Ackerman, "Study Links Mercury from Power Plants to Autism," *Houston Chronicle,* March 18, 2005. The original study: Raymond Palmer, Steven Blanchard, Zachary Stein, David Mandell, and Claudia Miller, "Environmental Mercury Release, Special Education Rates, and Autism Disorder: An Ecological Study of Texas," University of Texas Health and Science Center, November 1, 2004.

## 5 | The Autism Epidemic: Stealing Lives Around the World


1. John Hanchette, "Autism Statistics: Precipitous Increase in Autism Cases May Be Tied to Childhood Vaccines," *Niagara Falls Reporter,* February 24, 2004.
2. Autism statistics from the National Autistic Society, February 23, 2004.
3. "New Internal Documents Reveal Deception by the Centers for Disease Control about Vaccine's Role in Autism, Says Generation Rescue," U.S. Newswire, April 5, 2006.
4. Statistics taken from official state statistics produced by the Department of Education, "Autism Increases in the United States," the National Autism Association.
5. F. Edward Yazbak, M.D., "Autism in the United States: A Perspective," *Journal of American Physicians and Surgeons,* vol. 8, no. 4 (Winter 2003).
6. Hanchette, "Autism Statistics."
7. It was not until I came across the courageous investigation by Robert F. Kennedy, Jr., that I began to understand what had actually happened. His extensive article "Deadly Immunity," published in *Rolling Stone* and *Salon.com* in June 2005, was the result of even more extensive research that Kennedy conducted independently after being approached by the parents of autistic children. At first he was skeptical, but ultimately he became convinced that a cover-up of massive proportions had taken place and was still going on.
8. This concept was discussed in a September 13, 2005, *Op-EdNews.com* article, "FDA Knew Dangers of Thimerosal Vaccines for 60 Years," by Evelyn Pringle, an investigative journalist focused on exposing corruption in government.
9. Dr. Tim O'Shea, *The Sanctity of Human Blood: Vaccination Is Not Immunization,* Denver, Colorado: NewWest, 2001.
10. Quoted from a statement given to the Institute of Medicine Immunization Safety Committee of the National Academy of Sciences on January 11, 2001, by Barbara Loe Fisher, president and cofounder of the National Vaccine Information Center.
11. Barbara Loe Fisher's statement, "In the Wake of Vaccines," *Mothering Magazine,* September/October 2004.
12. "There's Mercury in Vaccines?" PutChildrenFirst.org.
13. "Simpsonwood," PutChildrenFirst.org.

14. Basic information from Robert F. Kennedy, Jr., "Deadly Immunity," as posted on *Salon.com,* June 16, 2005.
15. David Kirby, *Evidence of Harm,* page 4, New York: St. Martin's Press, 2005.
16. Evelyn Pringle, "Pharma's Poisoned Generation," *The Sierra Times,* November 29, 2005.
17. Dr. Mark A. Sircus, "Mercury, Vaccines and Medicine," International Medical Veritas Association.
18. Myron Levin, "CDC Not Opposed to Controversial Flu Shot," *Los Angeles Times,* April 3, 2004.
19. Kennedy, "Deadly Immunity."
20. Kirby, *Evidence of Harm,* prologue, pages 1–7.
21. Kennedy, "Deadly Immunity."
22. Ann Harding, "The Newest Vaccine," *Babytalk* magazine, October 2004.
23. Gardiner Harris and Anahad O'Connor, "On Autism's Cause, It's Parents vs. Research," *The New York Times,* June 25, 2005.
24. Kennedy, "Deadly Immunity."
25. Evelyn Pringle, "FDA Knew of Thimerosal Dangers 60 Years Ago," *Naples Sun Times,* April 27, 2005.
26. Kennedy, "Deadly Immunity."
27. Sallie Bernard, the Coalition for SafeMinds (www.safeminds.org), February 24, 2005.
28. Kennedy, "Deadly Immunity."

## 6| The Breast Cancer Industry: Mammograms, Ionizing Radiation, and Conflicts of Interest

1. Ruth Rosen, "Polluted Bodies," *San Francisco Chronicle,* February 3, 2003.
2. David Helwig, "Censoring Breast Cancer in San Francisco," *Canadian Medical Association Journal,* May 2, 2000.
3. "Tribute to Andrea Martin," Breast Cancer Fund (www.breastcancer.org), August 7, 2003.
4. "Breast Cancer Facts 2005: The Breast Cancer Epidemic," Breast Cancer Fund (www.breastcancer.org).
5. Nancy Evans and Jeanne Rizzo, "Can Rising Bay Area Breast Cancer Rates

Be Reversed?" *San Francisco Medical Society,* April 2005. Nancy Evans is a health science consultant for the Breast Cancer Fund. Jeanne Rizzo is its executive director.

6. "Chemical Industry Funds Breast Cancer Campaign," Cancer Prevention Coalition (www.preventcancer.com).
7. Sharon Batt and Lisa Gross, "Cancer, Inc.," *Sierra Magazine,* September/October 1999.
8. Samuel Epstein and Liza Gross, "The High Stakes of Cancer Prevention," *Tikkun* magazine, November/December 2000.
9. Epstein and Gross, "The High Stakes of Cancer Prevention."
10. "The Breast Cancer Epidemic 2005," Breast Cancer Fund (www.breastcancer.org).
11. "Politics of Breast Cancer," Breast Cancer Action (www.bcaction.org).
12. Samuel Epstein, "Cosmetics and Personal Care Products Can Be Cancer Risks," Cancer Prevention Coalition (www.preventcancer.com).
13. Epstein, "Cosmetics and Personal Care Products Can Be Cancer Risks."
14. Public health statement for 1,4 dioxane, Agency for Toxic Substances and Disease Registry, September 2004.
15. Epstein, "Cosmetics and Personal Care Products Can Be Cancer Risks."
16. David Steinman, "Hair Dyes to Die For," InterNatural (www.internatural-alternative-health.com), September 21, 2005.
17. "Study Examines Cancer Risk from Hair Dye," Reuters, CNN.com, January 28, 2004.
18. Epstein, "Cosmetics and Personal Care Products Can Be Cancer Risks."
19. Leah Hennen, "Skin Deep: Rocking the Baby Products Cradle," *The New York Times,* September 8, 2005.
20. Hennen, "Skin Deep."
21. Shelly Alpern, "Breast Cancer Activists Target Avon Products," Trillium Asset Management, a social investment company (www.trilliuminvest.com), August 2002.
22. Shelley Page, "Think Before You Pink," Breast Cancer Fund (www.breastcancerfund.org), April 17, 2005.
23. Page, "Think Before You Pink."
24. Page, "Think Before You Pink."
25. Page, "Think Before You Pink."
26. Samuel Epstein, Rosalie Bertell, and Barbara Seaman, "Dangers and Unreliability of Mammography: Breast Examination Is a Safe, Effective, and Practical

Alternative," Cancer Prevention Coalition (www.preventcancer.com), from an article in the *International Journal of Health Services,* 2001.

27. Epstein et al., "Dangers and Unreliability of Mammography."
28. Michael Moss, "Mammogram Team Learns from Its Errors," *The New York Times,* June 28, 2002.
29. Dawn Prate, "Mammograms Cause Breast Cancer (and Other Cancer Facts You Probably Never Knew)," NewsTarget.com, August 15, 2005.
30. Prate, "Mammograms Cause Breast Cancer."
31. Nancy Evans, *State of the Evidence: 2004,* Breast Cancer Fund.
32. Overview of the Long Island Breast Cancer Study Project, National Cancer Institute.
33. Janette Sherman, M.D., "The Breast Cancer Epidemic on Long Island," *Life's Delicate Balance: Causes and Prevention of Breast Cancer,* Oxford, England: Taylor & Francis, 2000.
34. Karl Grossman, "Long Island Breast Cancer Study Missing the Nuke Connection," *The East Hampton Star,* August 22, 2002.
35. Grossman, "Long Island Breast Cancer Study Missing the Nuke Connection."
36. Robert Ryan, "Cancer Research—A Super Fraud?" Campaign Against Fraudulent Medical Research (www.rense.com), 1997.
37. Batt and Gross, "Cancer, Inc."
38. Press release for Samuel S. Epstein, M.D., *Cancer-Gate: How to Win the Losing Cancer War.* Amityville, New York: Baywood, 2005.
39. Sherman, *Life's Delicate Balance.*

## 7| The New Lung Cancer Pandemic: Third-World Children, Contaminating Nonsmokers, and the Special Risks to Women

1. "Smoking and Tobacco," tobacco information, Centers for Disease Control and Prevention.
2. "Altria/Philip Morris Shareholders Meeting Talking Points," Global Solidarity Against Big Tobacco (www.essentialaction.org), April 23, 2003.
3. "Tobacco Marketing to Young People," INFACT's Tobacco Industry Campaign (www.infact.org).
4. Ross Hammond and Andy Rowell, "Trust Us: We're the Tobacco Industry," Action on Smoking and Health (www.ash.org.uk), May 2001.

5. "Tobacco Marketing to Young People."
6. "Tobacco Marketing to Young People."
7. "Global Solidarity Against Big Tobacco," Essential Information (www.essentialaction.org).
8. Clive Bates, Dr. Martin Jarvis, and Dr. Gregory Connolly, "ASH(UK) Report: Tobacco Additives—Cigarette Engineering and Nicotine Addiction," Action on Smoking and Health (ASH), July 14, 1999.
9. "Tobacco Marketing to Young People."
10. Dr. Elizabeth Fontham, "Environmental Tobacco Smoke and Lung Cancer in Nonsmoking Women," *Journal of Occupational and Environmental Medicine,* January 26, 1996.
11. "Smoking," "The Year in Medicine" issue, *Time* magazine, December 5, 2005.
12. Bates, Jarvis, and Connolly, "ASH(UK) Report."
13. Bates, Jarvis, and Connolly, "ASH(UK) Report."
14. Selected portions were taken from Elisa K. Ong and Stanton A. Glantz, "Constructing 'Sound Science' and 'Good Epidemiology': Tobacco Lawyers and Public Relation Firms," American Public Health Association, November 2000.
15. George Davey Smith and Andrew N. Phillips, "Passive Smoking and Health: Should We Believe Philip Morris's 'Experts'?" *British Medical Journal,* October 12, 1996.
16. Smith and Phillips, "Passive Smoking and Health."
17. "Nonsmokers with Lung Cancer," CBS News, August 10, 2005.
18. Olivia Barker, " 'Reeve's Death Leaves 'a Hole in the World,' " *USA Today,* March 7, 2006.
19. Janice Billingsley, "Women May Be More Susceptible to Lung Cancer Than Breast Cancer," *The Detroit News,* May 20, 2005.
20. Joseph W. Cherner, "Canadian Waitress with Secondhand Lung Cancer Fights for Life," *News Blaze,* February 28, 2006, parts excerpted from *Ottawa Citizen,* February 23, 2006.
21. "Smoking in Public Places," Forest (www.forestonline.org).
22. Glen Owen and Sarah Hills, "Tobacco Firms Beat Ban with 'Smoking Shelters,' " *The Mail,* April 9, 2006.
23. Thomas Mulier and Chris Burritt, "Tobacco Companies Keep Profiting Despite Regulation," *Bloomberg News,* January 2, 2006.
24. Dr. Ralph Moss, Ph.D. "Dr. Ralph Moss on Conventional Cancer Therapies," Healingdaily.com.

25. Dr. Samuel Epstein, "American Cancer Society: The World's Wealthiest 'Nonprofit' Institution," *International Journal of Health Services,* Vol. 29, no. 3 (1999).
26. Dr. Samuel Epstein et al., "The Crisis in U.S. and International Cancer Policy," *International Journal of Health Services,* Vol. 32, no. 4, (2002).
27. "The Inhalers," *Common Cause Magazine,* spring 1995.
28. Molly Ivins, "Molly Ivins on Tobacco Bill Death," *Fort Worth Star-Telegram,* June 21, 1998.
29. Robert Dreyfuss, excerpts from "George W. Bush: Calling For Philip Morris," *The Nation,* November 8, 1999.
30. Bill Straub, "Bush Nominates Tobacco Sympathizers to Key Posts," *The Detroit News,* January 5, 2001.
31. "Tobacco Sellout," *The Washington Post,* June 10, 2005.
32. "Tobacco Sellout."
33. "Anti-smoking Campaigner Dying of Lung Cancer," *Maclean's,* February 24, 2003, posted at *The Canadian Encyclopedia* (www.canadianencyclopedia.ca).
34. Justin Thompson, "Barbara Tarbox. Dying of Cancer, She Used Her Last Days to Warn Kids of the Dangers of Smoking," CBC News, May 19, 2003.
35. Thompson, "Barbara Tarbox."
36. Andrew Bridges, "Tobacco Will Kill 1 Billion This Century, Officials Say," Associated Press, *Common Dreams News Center,* June 11, 2006.

## 8| Saving the Children: A Spiritual Crisis, a Theological Opportunity

1. *Journal of Pharmacology and Experimental Therapeutics* study in ASH newsletter, May/June 1998.
2. "Sudden Infant Death Syndrome (SIDS)," *The Daily Progress,* Charlottesville, Virginia, October 12, 2005.
3. Bill Moyers, "9/11 and the Sport of God," *TomPaine*, September 9, 2005.
4. Glenn Scherer, "George Bush's War on Nature," *Salon.com,* January 6, 2003.
5. Glenn Scherer, "The Godly Must Be Crazy," *Grist* (www.gristmill.grist.org), October 27, 2004.
6. Scherer, "The Godly Must Be Crazy."
7. Bill Moyers, "There Is No Tomorrow," *Star Tribune,* January 30, 2005.

8. Michael Luo, "Doomsday: The Latest Word If Not the Last," *The New York Times,* October 16, 2005.
9. Scherer, "The Godly Must Be Crazy."
10. Scherer, "The Godly Must Be Crazy."
11. "Tom DeLay," Wikipedia (www.wikipedia.org), May 7, 2006.
12. Peter Perl, "Absolute Truth," *The Washington Post,* May 13, 2001.
13. Scherer, "The Godly Must Be Crazy."
14. Core material comes from Scherer, "The Godly Must Be Crazy," interview of James Inhofe by *Pentacostal Evangel* magazine.
15. Scherer, "The Godly Must Be Crazy."
16. Blaine Harden, "The Greening of Evangelicals," *The Washington Post,* February 6, 2005.
17. Harden, "The Greening of Evangelicals," quote from *Christianity Today,* Fall 2005.
18. "American Baptist Environmental statement," reprinted from the Acton Institute for the Study of Religion and Liberty at www.thegoodsteward.com, February 1, 2001.
19. Rabbi Saul Berman, "Jewish Environmental Values: The Dynamic Tension Between Nature and Human Needs," Jewish Virtual Library, 2005.
20. Daniel Kurtzman, " 'Green' Jews Gather Force in Environmental Movement," Jewish Telegraphic Agency, March 19, 1999.
21. Thich Tri Quang, "Environmental Protection," *Giac Ngo* magazine (www.saigon.com), July 1996.
22. Quoted in "An Invitation to Reflection and Action on Environment in Light of Catholic Social Teaching," U.S. Catholic Conference of Catholic Bishops, November 14, 1991.
23. "An Invitation to Reflection and Action on Environment in Light of Catholic Social Teaching."
24. Lara Lutz, "Environmental Stewardship Growing Presence in Churches," *Bay Journal,* February 2004.
25. Peter Sawtell, "Celebrating Movements, Taking Action," *Eco-Justice Notes* (newsletter of Eco-Justice Ministries, www.eco-justice.org), January 14, 2005.
26. Rachel Carson, *Silent Spring,* pages 242–43, New York: Mariner Books, 2002.
27. Robert Schlesinger, "Ecology Movement Now Getting Religion," *Free Republic* (www.freerepublic.com), July 8, 2001.
28. Blaine Harden, "'God's Mandate': Putting the White House on Notice," *The Washington Post,* February 6, 2005.

29. Rebecca Clarren, "Dirty Politics, Foul Air," *The Nation,* February 25, 2005.
30. Clarren, "Dirty Politics, Foul Air."
31. National Council of Churches, "Theologians Warn of 'False Gospel' on the Environment," *NCC News,* February 14, 2005.
32. National Council of Churches, "Theologians Warn of 'False Gospel' on the Environment."
33. Harden, "God's Mandate."
34. Geoffrey Lean, "Pollution Is a Sin, Say Church Leaders at Sea," Common Dreams News Center, June 9, 2002.
35. Scherer, "The Godly Must Be Crazy."

# INDEX